T0213592

Big and Integrated Artificial Intelligence

Volume 1

Series Editor

Vincenzo Loia, Department of Management and Innovation Systems, University of Salerno, Fisciano, Italy

The series "Big and Integrated Artificial Intelligence" (BINARI) publishes new developments and advances in the theory and computational aspects of AI in the context of Big Data Engineering. The aim is to address the issues related to the integration of different technologies, as well as the engineering aspects of the deployment of integrated AI technologies in real-world scenarios—quickly and with a high quality.

The intent is to cover the theory, applications, and design methods of BigAI as a dual concept of big data and AI, embedded in the fields of engineering, computer science, physics and life sciences, as well as the methodologies behind them. The series contains monographs, lecture notes and edited volumes in the Big and Integrated Artificial Intelligence intending to integrate multiple AI technologies (e.g., vision, speech, real-time control) in order to develop efficient and robust systems interacting in the real world.

Topics covered include, but are not limited to, multimodal representations and modeling, deep architectures and learning algorithms, self-monitoring systems, high-dimensional data driven systems, optimization schemes, uncertainty handling.

Of particular value to both the contributors and the readership are the short publication timeframe and the world-wide distribution, which enable both wide and rapid dissemination of research output.

More information about this series at https://link.springer.com/bookseries/16396

Stefania Tomasiello · Witold Pedrycz · Vincenzo Loia

Contemporary Fuzzy Logic

A Perspective of Fuzzy Logic with Scilab

 Springer

Stefania Tomasiello
Institute of Computer Science
University of Tartu
Tartu, Estonia

Witold Pedrycz
Department of Electrical and Computer
Engineering
University of Alberta
Edmonton, AB, Canada

Vincenzo Loia
Department of Management and Innovation
Systems
University of Salerno
Fisciano, Salerno, Italy

ISSN 2662-4133   ISSN 2662-4141   (electronic)
Big and Integrated Artificial Intelligence
ISBN 978-3-030-98976-7   ISBN 978-3-030-98974-3   (eBook)
https://doi.org/10.1007/978-3-030-98974-3

This Springer imprint is published by the registered company Springer Nature Switzerland AG
The registered company address is: Gewerbestrasse 11, 6330 Cham, Switzerland

To all the enthusiastic students and learners

Preface

Writing a book is like a journey through past experiences that one may want to share with potential readers. The Italian writer Italo Calvino in "Six Memos for the Next Millennium", a book of undelivered lectures scheduled at Harvard University, pinpoints the essential features of a book: lightness, quickness, exactitude, visibility, multiplicity, and consistency. Even though these features were meant for novels, we believe that most of them also apply to a scientific textbook. Armed with such a vision, we wish to offer to the readers a synthesis of our research and experience through the courses we have delivered to upper-level undergraduate students and to graduate students majoring in computer science, mathematics, physics, and engineering.

More than half a century has passed since fuzzy logic appeared, but the research in this area is still very active, with many applications in different fields of our modern society. The COVID-19 pandemic, which has upset everyone's life, has also brought changes in our work habits. More than ever, especially for students, the availability of free and open-source software has been important.

In this book, the reader is introduced to basic concepts that span the classical notions of fuzzy logic to more advanced notions from the current state-of-the-art research. Each of the major topics is accompanied with examples and Scilab codes. To the best of our knowledge, this is the first book presenting topics in fuzzy logic with the support of free open-source software, such as Scilab. Even though this book may be used as a textbook for some courses, there are sufficient ideas for starting research projects in fuzzy logic.

We wish to thank our students, whose questions inspired many of this book's examples and problems. Stefania Tomasiello acknowledges funding from the European Social Fund via the IT Academy programme.

Tartu, Estonia

Edmonton, Canada

Salerno, Italy

Stefania Tomasiello

Witold Pedrycz

Vincenzo Loia

Contents

Chapter 1
Introduction to Fuzzy Sets

1.1 Fuzzy Sets: Some Definitions

The fuzzy set theory was introduced in 1965 by Lotfi Asker Zadeh, an electrical engineer and researcher in mathematics, computer science and pioneer of artificial intelligence. This theory was born with the aim to impart a mathematical treatment on certain subjective terms of language, such as "about" and "around", among others, and to perform calculations on vague or flexible entities, as human beings do.

More often than not, the classes of objects encountered in the real physical world do not have precisely defined criteria of membership. For example, the class of animals clearly includes dogs, horses, birds, etc. as its members and clearly excludes such objects as rocks, fluids, plants, etc. However, such objects as starfish, bacteria, etc. have an ambiguous status with respect to the class of animals. [...] Clearly, the "class of all real numbers which are much greater than 1," or "the class of beautiful women" , or "the class of tall men" do not constitute classes or sets in the usual mathematical sense of these terms. Yet, the fact remains that such imprecisely defined "classes" play an important role in human thinking, particularly in the domains of pattern recognition, communication of information, and abstraction (Zadeh 1965).

The formal mathematical representation of a fuzzy set is based on the fact that any set can be characterized by a function, i.e. its characteristic function. Its counterpart in the fuzzy set theory is the membership function.

Most real systems are not deterministic and they cannot be described precisely. For instance, one can easily realize this by looking at stock market, smart grids with renewable energy sources, road traffic.

The term fuzziness is used to indicate the vagueness concerning the description of the semantic meaning of the events, phenomena, or statements themselves; formally, the degree to which the element satisfies properties characterized by a fuzzy set. As an expression of the possibility theory (Zadeh 1978), it represents a persisting uncertainty in conditions where the event remains unclear even after its occurrence, unlike the probability theory which formalizes uncertainty in situations with given boundaries.

© The Author(s), under exclusive license to Springer Nature Switzerland AG 2022
S. Tomasiello et al., *Contemporary Fuzzy Logic*, Big and Integrated
Artificial Intelligence 1, https://doi.org/10.1007/978-3-030-98974-3_1

1.1.1 Sets and Fuzzy Sets

Let X be a space of points (objects), with a generic element of X denoted by x.

A set is normally defined as a collection of elements or objects $x \in X$ that can be finite, countable. Any element x may belong or not to a set A, $A \subset X$.

The member elements of A can be defined by means of the characteristic function $\chi_A : X \to \{0, 1\}$, which uses 1 to indicate membership and 0 nonmembership. For a fuzzy set, the membership function allows various degrees of membership for the elements of a given set.

Definition 1.1 A *fuzzy set* A can be regarded as a set of ordered pairs:

$$A = \{(x, \mu_A(x)) | x \in X\}, \tag{1.1}$$

where

$$\mu_A : X \to [0, 1] \tag{1.2}$$

is the membership function of A and $\mu_A(x)$ is the degree of membership of x in A.

The value $\mu(x) = 0$ indicates complete nonmembership of x in A, while the value $\mu(x) = 1$ represents complete membership of x in A; values in between are used to represent intermediate degrees of membership. X is usually referred to as the *universe of discourse* or simply *universe*. It may be a discrete set or an interval of the real line. The first case applies to discrete or finite fuzzy sets, while the second case holds for continuous fuzzy sets. A continuous fuzzy set is uniquely represented by its membership function (MF).

The membership function is fixed according to the experience or domain knowledge. It is possible to distinguish three approaches in this regard:

- psychological approach, when some experts in the application field of the case study are questioned about the most suitable membership curve for the given problem;
- statistical approach, when some data in the form of frequency histograms or other probability curves are used to construct a membership function;
- automatic approach, usually based on neural networks and/or genetic algorithms, and adopted when no expert is available or when there are so many data about the case study that they can be automatically processed (Fig. 1.1).

A *finite* fuzzy set can also be represented as follows

$$A = \sum_{i=0}^{n} \frac{\mu(x_i)}{x_i}. \tag{1.3}$$

The notation $\frac{\mu(x_i)}{x_i}$ does not indicate division, it is a way to visualize an element x_i and its respective degree of membership μ_A. Similarly, $\sum$ does not mean summation, it is a way to connect the elements of X that are in A with their respective degrees.

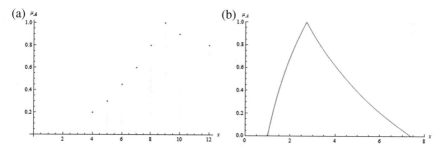

Fig. 1.1 Fuzzy sets: **a** finite, **b** continuous

Let A and B be two continuous fuzzy sets, with their membership functions $\mu_A(x)$, $\mu_B(x)$, with $x \in X$. Then A and B are said to be equal if and only if

$$\mu_A(x) = \mu_B(x), \quad \forall x \in X. \tag{1.4}$$

If A and B are two discrete fuzzy sets, then

$$\frac{\mu_A(x_i)}{x_i} = \frac{\mu_B(x_i)}{x_i}, \quad \forall x_i \in X. \tag{1.5}$$

A fuzzy set that has only one point $\xi \in X$ with $\mu_A(\xi) = 1$ is called a *singleton*.

Example 1.1 Consider the set of natural prime numbers:

$$P = \{p \in \mathbb{N} : p \ \ is \ \ prime\}.$$

The characteristic function $\chi_P(p)$ is

$$\chi_P = \begin{cases} 1, if \ \ p \ \ is \ \ prime \\ 0, \quad \quad \ otherwise. \end{cases}$$

Example 1.2 Consider the subset of real numbers "close to 3":

$$A = \{x \in \mathbb{R} : x \ \ is \ \ close \ \ to \ \ 3\}.$$

Do the numbers 6 and 3.01 belong to A? It is debatable to objectively say when a number is close to 3. It is reasonable to say that 3.01 is nearer 3 than 6.

We can define a function $\mu_A : \mathbb{R} \to [0, 1]$ as follows (Fig. 1.2)

$$\mu_A = \begin{cases} 1 - |x - 3|, if \ \ 2 < x < 4 \\ 0, \quad \ otherwise. \end{cases} \tag{1.6}$$

Fig. 1.2 Example 1.2:
membership function $\mu_A(x)$

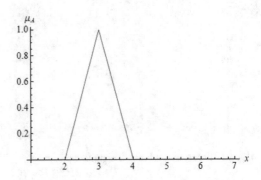

So it is:
$\mu_A(3.01) = 0.99$, $\mu_A(3.5) = 0.5$ and $\mu_A(6) = 0$.
If the closeness (proximity) function is defined by

$$\nu_A(x) = \exp[-(x-3)^2],\qquad\qquad(1.7)$$

with $x \in \mathbb{R}$, then the degrees of membership change as follows:

$$\nu_A(3.01) = 0.9999,\quad \nu_A(3.5) = 0.778801,\quad \nu_A(6) = 0.00012341.$$

The choice of the membership function is subjective, although it has to reflect the semantics of the concept described by the fuzzy set. In Fig. 1.3 an alternative membership function is depicted.

It is worth mentioning that the concept "numbers close to 3" by a classical set can be expressed by considering a sufficiently small value of ϵ and the characteristic function for the interval $(3 - \epsilon, 3 + \epsilon)$ as follows

$$\chi_A = \begin{cases} 1, & if \quad |x-3| < \epsilon \\ 0, & if \quad |x-3| > \epsilon. \end{cases}$$

Fig. 1.3 Example 1.2:
alternative membership
function $\nu_A(x)$

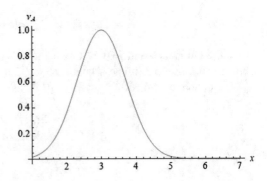

Fig. 1.4 Fuzzy sets from Example 1.3

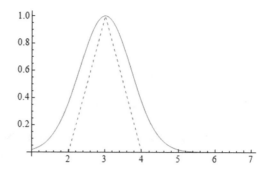

Then being close to 3 means being within a certain neighborhood of 3. In this case, it is the choice of the radius of the neighborhood considered to be subjective, but all the values within the neighborhood are close to 3 with the same degree of membership, which is 1.

Let A and B be two fuzzy sets, with their respective membership functions μ_A and μ_B.

Definition 1.2 A is a fuzzy subset of B, and write $A \subseteq B$, if $\mu_A(x) \le \mu_B(x)$ for all $x \in X$.

Example 1.3 Let A and B be two fuzzy sets defined by the membership functions (1.6) and (1.7) respectively. Then $A \subset B$ (see Fig. 1.4).

1.1.2 Support, Core, Height

Three important concepts are support, core and height of a fuzzy set.

The *support* of a fuzzy set A is the set of all points $x \in X$ whose membership degree in A is positive:

$$supp(A) = \{x | \mu_A(x) > 0\}. \tag{1.8}$$

The *core* of a fuzzy set A is the set of all points $x \in X$ whose membership degree in A equals 1:

$$core(A) = \{x | \mu_A(x) = 1\}. \tag{1.9}$$

The height of a fuzzy set A in X is defined as

$$hgt(A) = \max_{x \in X} \mu_A(x). \tag{1.10}$$

A fuzzy set A is *normal* if its height is equal to 1 (Fig. 1.5). Otherwise, A is called subnormal fuzzy set.

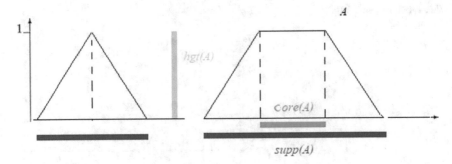

Fig. 1.5 Triangular (on the left) and trapezoidal (on the right) membership functions and their core, height and support

A fuzzy set A is *convex* if and only if for any $x_1, x_2 \in X$ and any $\lambda \in (0, 1)$,

$$\mu_A(x_3) \geq \min\{\mu_A(x_1), \mu_A(x_2)\},\tag{1.11}$$

with $x_3 = \lambda x_1 + (1 - \lambda)x_2$.

If (1.11) is a strict inequality, then the fuzzy set is said to be strongly convex. In Fig. 1.6a, a convex fuzzy set is depicted. The plotmarkers "+", "−" and "*" represent the points obtained at the abscissae x_1, x_2, x_3. The continuous convex fuzzy set is represented by a dotted line.

One can prove that if A is a convex fuzzy set, then its core is a convex set. Besides, if A and B are convex, then their intersection is convex.

A fuzzy set which satisfies normality and convexity represents a fuzzy number.

Example 1.4 This example's purpose is to model as a fuzzy set the set of "tall men". Considering a man tall depends on several factors such as country, observer point of view. For instance, in Bosnia and Herzegovina, Denmark and Estonia, the average

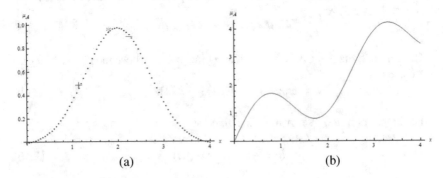

Fig. 1.6 a A convex fuzzy set. **b** A non-convex fuzzy set

Fig. 1.7 Fuzzy set of tall men

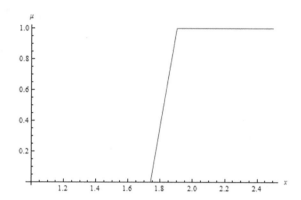

height of men is 1.82 m (https://www.worlddata.info/average-bodyheight.php#by-population). Let us suppose that the universe is $X = [0, 2.5]$, where each element represents the height in meter. A possible membership function for the fuzzy set tall men is (Fig. 1.7)

$$\mu_A = \begin{cases} (x - 1.77)/0.1, & if \quad 1.77 \leq x \leq 1.87 \\ 1, & if \quad x > 1.87 \\ 0, & otherwise. \end{cases}$$

For this fuzzy set, the support is $supp(A) = (1.77, 2.5]$; the core is $core(A) = [1.87, 2.50]$, the height is $hgt(A) = 1$

Example 1.5 Consider the universe $X = [0, 400]$, where $x \in X$ represents the area of a flat in m². To model the fuzzy set of "large flats" one may consider the fuzzy set with the following membership function

$$\mu_A = \begin{cases} 0, & if \quad 0 \leq x \leq 90 \\ \frac{x-90}{90}, & if \quad 90 < x < 180 \\ 1, & if \quad 180 \leq x \leq 400. \end{cases}$$

1.2 Operations on Fuzzy Sets

Let A and B be two fuzzy sets, with their membership functions $\mu_A(x)$, $\mu_B(x)$.

Definition 1.3 The union $C = A \cup B$ is the fuzzy set whose membership function is:

$$\mu_C(x) = \max\{\mu_A(x), \mu_B(x)\}, \quad x \in X. \tag{1.12}$$

This represents an extension of the classical case. In fact, if A and B were sets:

$$\max\{\chi_A(x), \chi_B(x)\} = \begin{cases} 1, & if \quad x \in A \quad or \quad x \in B \\ 0, & if \quad x \notin A \quad and \quad x \notin B \end{cases}$$

$$= \begin{cases} 1, & if \quad x \in A \cup B \\ 0, & if \quad x \notin A \cup B \end{cases}$$

$$= \chi_{A \cup B}(x), \quad x \in X.$$

Definition 1.4 The intersection $D = A \cap B$ is the fuzzy set whose membership function is:

$$\mu_D(x) = \min\{\mu_A(x), \mu_B(x)\}, \qquad x \in X. \tag{1.13}$$

In case of classical sets:

$$\min\{\chi_A(x), \chi_B(x)\} = \begin{cases} 1, & if \quad x \in A \quad and \quad x \in B \\ 0, & if \quad x \notin A \quad or \quad x \notin B \end{cases}$$

$$= \begin{cases} 1, & if \quad x \in A \cap B \\ 0, & if \quad x \notin A \cap B \end{cases}$$

$$= \chi_{A \cap B}(x), \quad x \in X.$$

Definition 1.5 The complement of A is the fuzzy set A' whose membership function is given by

$$\mu_{A'} = 1 - \mu_A(x), \qquad x \in X. \tag{1.14}$$

Notice that, unlike the classical situation, it can be

$$\mu_{A \cap A'}(x) \neq 0, \qquad \mu_{A \cup A'}(x) \neq 1. \tag{1.15}$$

An example of all the above-mentioned operations on two fuzzy sets is offered in Fig. 1.8.

Example 1.6 Let us recall Examples 1.4 and 1.5. The fuzzy sets of "not tall men" and "small flats" have the membership function $\mu_{A'} = 1 - \mu_A$, representing the complement of the mentioned fuzzy sets. Notice that the intersection of A and A' is nonempty in both examples.

Notice that while $\mu_A(x)$ represents the degree of compatibility of x with a certain linguistic concept, $\mu_{A'}(x)$ shows the incompatibility of x with the same concept. As a consequence of the imprecision of fuzzy sets, there is a certain overlap of a fuzzy set with its complement.

The operations between fuzzy sets satisfy the following properties:

- $A \cup B = B \cup A$ and $A \cap B = B \cap A$ (commutativity),
- $A \cup (B \cup C) = (A \cup B) \cup C$ and $A \cap (B \cap C) = (A \cap B) \cap C$ (associativity),
- $A \cup A = A$ and $A \cap A = A$ (idempotency),
- $A \cup (B \cap C) = (A \cup B) \cap (A \cup C)$ and $A \cap (B \cup C) = (A \cap B) \cup (A \cap C)$ (distributivity),
- $A \cap \emptyset = \emptyset$ and $A \cup \emptyset = A$ (identity),
- $(A \cup B)' = A' \cap B'$ and $(A \cap B)' = A' \cup B'$ (De Morgan's law)

The Scilab function `ffs=fintersect(set1,md1,set2,md2)` returns the intersection of two finite fuzzy sets `set1`, `set2`, given as row vectors collecting their universe elements, once their membership degrees `md1`, `md2` are also provided as row vectors. The code first check whether `set1`, `set2` have common elements:
`elements = intersect(set1,set2);`

```
if elements==[] then error('no intersection');
else
membership = zeros(1,length(elements));
```
then it finds iteratively the membership grades of the common elements (if any) by the `min` command
```
for i = 1:length(elements)
membership(i) = min(md1(set1==elements(i)),
md2(set2==elements(i)));
end
```

The Scilab function `ffs=funion(set1,md1,set2,md2)` to find the union of two finite fuzzy sets has the same arguments as the previous function for the intersection. The code has to take into account both possible common elements and different elements (the latter case is the simplest one). Then the membership grades are iteratively found by the `max` command.

The complete code listings for the union and intersection of finite fuzzy sets is provided in the Appendix.

Example 1.7 Consider two finite fuzzy sets (Fig. 1.9)

$$A = \{(4, 0.2), (5, 0.3), (6, 0.4), (7, 0.6), (8, 0.8), (9, 1), (10, 1), (12, 0.8)\},$$

$$B = \{(5, 0.03), (6, 0.06), (7, 0.1), (8, 0.2), (9, 0.5), (10, 1.), (11, 0.5)\}.$$

Their support is:
$supp(A) = \{4, 5, 6, 7, 8, 9, 10, 12\}, \quad supp(B) = \{5, 6, 7, 8, 9, 10, 11\}.$
Their core is:
$core(A) = \{9, 10\}, \quad core(B) = \{10\}.$ Their height is 1.
The intersection and the union of the two given fuzzy sets are respectively:

$$A \cap B = \{(5, 0.03), (6, 0.06), (7, 0.1), (8, 0.2), (9, 0.5), (10, 1)\},$$

$$A \cup B = \{(4, 0.2), (5, 0.3), (6, 0.4), (7, 0.6), (8, 0.8), (9, 1),$$

To run this example in Scilab, mind that the two given finite fuzzy sets can be represented as matrices:

```
A = [4,0.2;5,0.3; 6,0.4;7,0.6;8,0.8;9,1;10,1;12,0.8];
B = [5,0.03;6,0.06;7,0.1;8,0.2;9,0.5;10,1;11,0.5];
```

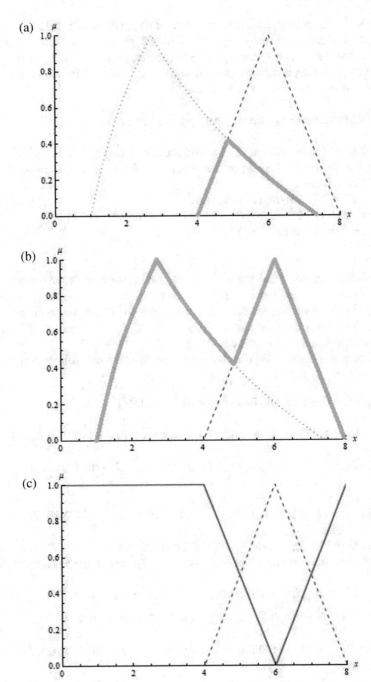

Fig. 1.8 Operations on fuzzy sets **a** union **b** intersection, and **c** complement (all represented by a grey thick line)

Fig. 1.9 Example
1.7—Finite fuzzy sets A
(blue) and B (red)

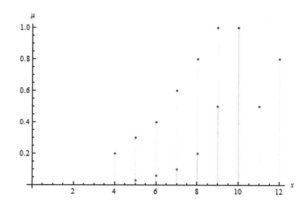

The elements of the fuzzy sets can be collected in two vectors
```
fs1=A(:,1)';
fs2=B(:,1)';
```
as their membership grades
```
md1=A(:,2)';
md2=B(:,2)';
```
The needed functions have to be executed:
```
exec('chs.sce',-1)
exec('fintersect.sce',-1)
exec('funion.sce',-1)
```
and invoked to get core, height, support of the two fuzzy sets:
```
[core_A,height_A,supp_A]=chs(fs1,md1);
[core_B,height_B,supp_B]=chs(fs2,md2);
```
along with intersection and union, respectively:
```
AintersectB=fintersect(fs1,md1,fs2,md2);
AunionB=funion(fs1,md1,fs2,md2);
```
The output is visualized by the line command
```
disp(AunionB,"AunionB",AintersectB,"AintersectB",
core_B,"core_B",height_B,"height_B",supp_B,"supp_B",
core_A,"core_A",height_A,"height_A",supp_A,"supp_A")
```

Fig. 1.10 shows the screenshot of the console with the results.

Example 1.8 Let $X = \{1, 2, \ldots, 10\}$ be the universe. Consider the finite fuzzy sets:

$$A = \{(2, 0.4), (3, 0.6), (4, 0.8), (5, 1), (6, 0.8), (7, 0.6), (8, 0.4)\},$$

$$B = \{(2, 0.4), (4, 0.8), (5, 1), (7, 0.6)\}.$$

The cardinality and relative cardinality of B is $|B| = 0.4 + 0.8 + 1 + 0.6 = 2.8$, $\|B\| = 2.8/10 = 0.28$

```
Scilab 6.0.2 Console

--> exec('D:\codes\ch1\example1_7.sci', -1)

 supp_A

   4.    5.    6.    7.    8.    9.    10.    12.

 height_A

   1.

 core_A

   9.    10.

 supp_B

   5.    6.    7.    8.    9.    10.    11.

 height_B

   1.

 core_B

   10.

 AintersectB

   5.      0.03
   6.      0.06
   7.      0.1
   8.      0.2
   9.      0.5
   10.     1.

 AunionB

   4.      0.2
   5.      0.3
   6.      0.4
   7.      0.6
   8.      0.8
   9.      1.
   10.     1.
   12.     0.8

-->
```

Fig. 1.10   Example 1.7: Scilab results

The intersection of A with its complement $A' = \{(2, 0.6), (3, 0.4), (4, 0.2), (5, 0), (6, 0.2), (7, 0.4), (8, 0.6)\}$, is

$$A \cap A' = \{(2, 0.4), (3, 0.4), (4, 0.2), (5, 0), (6, 0.2), (7, 0.4), (8, 0.4)\}.$$

1.3 α-Level Sets

Definition 1.6 The α-*level* of a fuzzy set A is the set A_α defined by

$$A_\alpha = \{x \in X | \mu_A(x) \geq \alpha\}, \qquad 0 < \alpha \leq 1. \tag{1.16}$$

For $\alpha = 0$, the α-level of A is

$$A_0 = cl(\{x \in X | \mu_A(x) > 0\}), \tag{1.17}$$

where cl denotes the closure in the standard topology of X.

It is possible to prove that given two fuzzy sets A and B, $A = B$ if and only if $A_\alpha = B_\alpha$ for all $\alpha \in [0, 1]$.

By using the definition of α-levels, the following properties hold

- $(A \cup B)_\alpha = A_\alpha \cup B_\alpha$
- $(A \cap B)_\alpha = A_\alpha \cap B_\alpha$

Remark 2. A fuzzy set A is normal if all its α-levels are not empty, that is $A_1 \neq \emptyset$.

Remark 3. A fuzzy set can be defined as

$$A = \bigcup_{\alpha \in [0,1]} A_\alpha,$$

where $\cup$ denotes the standard fuzzy union.

In Fig. 1.11 an α-level is depicted.

Example 1.9 Let A be the fuzzy set represented by

$$A = \frac{0.1}{1} + \frac{0.25}{2} + \frac{0.35}{3} + \frac{0.7}{5} + \frac{0.9}{8} + \frac{1}{10}.$$

Then

$$A' = \frac{0.9}{1} + \frac{0.75}{2} + \frac{0.65}{3} + \frac{0.3}{5} + \frac{0.1}{8} + \frac{0}{10}.$$

Fig. 1.11 α-level of the
fuzzy set A

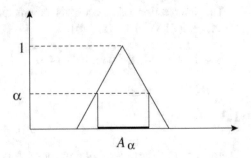

Some α-level sets are listed below:

$$A_{0.2} = \{2, 3, 5, 8, 10\}, \quad A'_{0.2} = \{1, 2, 3, 5\}$$

$$A_{0.5} = \{5, 8, 10\}, \quad A'_{0.5} = \{1, 2, 3\}$$

$$A_{0.8} = \{8, 10\}, \quad A'_{0.8} = \{1\}$$

$$A_1 = \{10\}$$

1.4   Generalizations of Fuzzy Sets

In this section we briefly recall some types of fuzzy sets, as extensions of fuzzy sets. The fuzzy sets presented so far can be regarded as type-1 fuzzy sets.

1.4.1   Type-2 Fuzzy Sets

Zadeh (1975) introduced the concept of a type-2 fuzzy set as an extension of an ordinary fuzzy set, i.e., a type-1 fuzzy set. Mizumoto and Tanaka (1976) studied the set theoretic operations of type-2 fuzzy sets and related properties.

Mendel and John (2002) provided further clarifications on the use of type-2 fuzzy sets.

Definition 1.7 Let $\mathcal{F}([0, 1])$ be the set of all ordinary fuzzy sets on $[0, 1]$. A *type-2 fuzzy set* is a fuzzy set in X, whose membership values are fuzzy sets on $[0, 1]$, $A : X \rightarrow \mathcal{F}([0, 1])$, and which can be represented as follows

$$A = \{((x, u), \mu_A(x, u)) | \forall x \in X, \forall u \in J_x \subset [0, 1]\}, \tag{1.18}$$

Fig. 1.12 A type-2 fuzzy set (LMF, dashed line; UMF, continuous line; FUO, grey area)

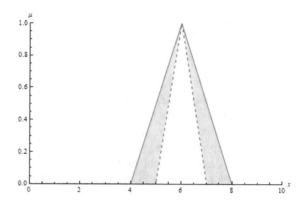

where x is the primary variable, J_x is the primary membership of x, u is the secondary variable.

Type-2 fuzzy sets can also be represented as follows

$$A = \frac{\int_{x \in X} f(x)}{x},$$ (1.19)

where $f(x) = \frac{\int_{u \in J_x} \mu(x,u)}{u}$ is the secondary membership function.

For $x = x'$, the 2D plane whose axes are u and $\mu_A(x', u)$ is called a vertical slice of $\mu_A(x, u)$ and it represents a secondary membership function. The type-2 fuzzy set can be regarded as three-dimensional, where the third dimension is the value of the membership function at each point on its two-dimensional domain that is called its *footprint of uncertainty* (FUO). The FOU is completely described by two bounding functions, namely a lower membership function (LMF) and an upper membership function (UMF), which are both classical (type-1) fuzzy sets (Fig. 1.12).

The type-2 fuzzy set can be generalized as follows.

Definition 1.8 A *type-m* fuzzy set is a fuzzy set in X whose membership values are type $m - 1$, with $m > 1$, fuzzy sets on $[0, 1]$.

1.4.2 Intuitionistic, Neutrosophic and Rough Sets

Definition 1.9 Given an underlying set X of objects, an *intuitionistic fuzzy set* (IFS) A is a set of ordered triples,

$$A = \{(x, \mu_A(x), \nu_A(x)) | x \in X\}$$ (1.20)

where $\mu_A(x) : X \to [0, 1]$ and $\nu_A(x) : X \to [0, 1]$ represent the degree of membership and the degree of nonmembership of the element x to A respectively, through $\pi_A(x) = 1 - \mu_A(x) - \nu_A(x)$.

The condition $0 \leq \mu_A(x) + \nu_A(x) \leq 1$ holds.

Ordinary fuzzy sets over X may be regarded as special intuitionistic fuzzy sets with the nonmembership function $\nu_A(x) = 1 - \mu_A(x)$.

In the *neutrosophic set* there is also a nonmembership value, but it is taken into account in a different way. Let T, I, F denote the membership, indeterminacy, and nonmembership values, respectively.

Let U be the universe of discourse, and M a set included in U. An element $x(T, I, F)$ belongs to M in the following way: it is $T\%$ true in the set, $I\%$ indeterminate (unknown if it is) in the set, and $F\%$ false.

For instance, $x(0.5, 0.2, 0.3)$ belongs to M means that with a probability of 50% x is in M, with a probability of 30% x is not in M, and the rest is undecidable.

Rough sets may represent a useful tool in processing incomplete and insufficient information. The starting point of rough set theory is the indiscernibility relation, which is generated from the information about objects of interest. This relation expresses the fact that it is not possible to discern some granules (or clusters) of objects, each one having the same properties, based on the available information (or knowledge).

Problems

1. Find a fuzzy set to model "high-quality item".
2. Elaborate on and define the corresponding universe of a discrete fuzzy set for the diagnosis of mind disorders.
3. With regard to the previous finite fuzzy set, find the height, core and support.
4. Consider a certain continuous fuzzy set. Show that the intersection of the fuzzy set with its complement is not null.

References

Mendel JM, John RI (2002) Type-2 fuzzy sets made simple. IEEE Trans Fuzzy Syst 10(2):117–127
Mizumoto M, Tanaka K (1976) Some properties of fuzzy sets of type-2. Inform Control 31:312–340
Zadeh LA (1965) Fuzzy sets. Inf Control 8:338–353
Zadeh LA (1975) The concept of linguistic variables and its application to approximate reasoning I, II, III. Inf Sci 8:199–249, 301–357, 43–80
Zadeh LA (1978) Fuzzy sets as a basis for a theory of possibility. Fuzzy Sets Syst 1:3

Chapter 2
Fuzzy Numbers

2.1 Zadeh's Extension Principle

Let $f : X \to Y$ be a real function. Besides, let A be a fuzzy set in X, with its membership function $\mu_A(x)$, and B a fuzzy set in Y defined as follows:

$$B = \{(y, \mu_B(y)) | y = f(x), x \in X\} \tag{2.1}$$

According to the Zadeh's extension principle (Zadeh 1975), a function $f : X \to Y$ induces another function $\overline{f} : \mathcal{F}(X) \to \mathcal{F}(Y)$, which represents the membership function of B:

$$\overline{f}(A)(y) = \mu_B(y) = \begin{cases} \sup\limits_{x \in f^{-1}(y)} \mu_A(x), & if \quad f^{-1}(y) \neq 0 \\ 0, & otherwise \end{cases} \tag{2.2}$$

where $f^{-1}(y) = \{x : f(x) = y\}$ is the preimage of y.

If f is a bijective function then f^{-1} is the inverse of f. Besides, if f is injective, then $y = f(x)$ belongs to B with the same degree α as x belongs to A. This may not happen if f is not injective.

The definition above can be extended to the multidimensional case. Let $X = X_1 \times \cdots \times X_r$ and $A_1, \ldots, A_r$ be r fuzzy sets in $X_1, \ldots, X_r$ respectively. Let f be a mapping from X to a universe Y, $y = f(x_1, \ldots, x_r)$. Then according to the extension principle, there is a fuzzy set B in Y

$$B = \{(y, \mu_B(y)) | y = f(x_1, \ldots, x_r), (x_1, \ldots, x_r) \in X\} \tag{2.3}$$

where

$$\mu_B(y) = \begin{cases} \sup\limits_{(x_1,\ldots,x_r) \in f^{-1}(y)} \min\{\mu_{A_1}(x_1), \ldots, \mu_{A_r}(x_r)\}, & if \quad f^{-1}(y) \neq 0 \\ 0, & otherwise. \end{cases} \tag{2.4}$$

S. Tomasiello et al., *Contemporary Fuzzy Logic*, Big and Integrated
Artificial Intelligence 1, https://doi.org/10.1007/978-3-030-98974-3_2

Fig. 2.1 Example 2.1:
extension principle for the
function $f(x) = \exp(x)$

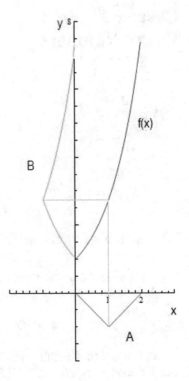

Example 2.1 Let $f(x) = \exp(x)$ and the fuzzy set A be defined by

$$\mu_A(x) = \begin{cases} x, & if \ \ 0 \le x < 1 \\ -x + 2, & if \ \ 1 \le x \le 2 \\ 0, & otherwise. \end{cases}$$

By applying the extension principle, one gets (Fig. 2.1)

$$\mu_B(y) = \begin{cases} log(y), & if \ \ 1 \le y < e \\ -log(y) + 2, & if \ \ e \le y \le e^2 \\ 0, & otherwise. \end{cases}$$

Example 2.2 Let $A = \{(-2, 0.4), (0, 0.8), (1, 0.6), (2, 0.5)\}$ and $f(x) = x^2$. First, we notice that by applying the extension principle, for $y = 4$, that is $x = \pm 2$, one gets

$$\mu_B(1) = \max_{\{x^2=4\}} \mu_A(x) = \max\{0.5, 0.4\}$$

and the resulting fuzzy set (i.e. considering $x = 0$ and $x = 1$, in addition to $x = \pm 2$) is

$$B = \{(0, 0.8), (1, 0.6), (4, 0.5)\}.$$

Example 2.3 Let $f : \mathbb{R} \times \mathbb{R} \to \mathbb{R}$ be a function that satisfies $f(x_1, x_2) = x_1 \times x_2$. Consider the finite fuzzy sets of $\mathbb{R}$

$$A_1 = 0.5/4 + 0.8/5 + 0.5/6 + 1/7,$$

$$A_2 = 0.5/7 + 1/8 + 0.5/9 + 0.2/10.$$

The membership degree of $y = f(x_1, x_2) = 40$ in B is

$$\mu_B(40) = \sup_{\{x_1 \times x_2 = 40\}} \min[\mu_{A_1}, \mu_{A_2}] = \tag{2.5}$$

$$= \max\{\min[\mu_{A_1}(4), \mu_{A_2}(10)], \min[\mu_{A_1}(5), \mu_{A_2}(8)]\} = \tag{2.6}$$

$$= \max\{0.2, 0.8\} = 0.8. \tag{2.7}$$

2.2 Fuzzy Numbers

A fuzzy number A over $\mathbb{R}$ satisfies the following conditions:

- all the α-levels of A are not empty for $0 \leq \alpha \leq 1$ (the core of A is not empty);
- all the α-levels of A are closed intervals of $\mathbb{R}$;
- $supp(A)$ is bounded.

The set of all fuzzy numbers will be denoted by $\mathcal{F}(\mathbb{R})$, and the set of the real numbers $\mathbb{R}$ is a subset of $\mathcal{F}(\mathbb{R})$.

Definition 2.1 A fuzzy number A is called positive (negative) if its membership function is such that $\mu_A(x) = 0, \forall x < 0$ ($\forall x > 0$).

Remark 2.1 Every real number r is a fuzzy number whose membership function is the characteristic function:

$$\chi_r(x) = \begin{cases} 1, & if \quad x = r \\ 0, & otherwise \end{cases} \tag{2.8}$$

and it is denoted by χ_r or $\tilde{r}$.

2.2.1 Parametric Form of Fuzzy Numbers and Types of Fuzzy Numbers

The parametric form (or α-cut representation) of fuzzy numbers uses the lower and upper extremal values of the α-cuts. Since an α-cut is an interval, arithmetic operations can be performed by using the expressions of the lower and upper boundary of each α-cut.

Definition 2.2 The *parametric form* of the fuzzy number A is given by a pair $\left[\underline{a}(\alpha), \overline{a}(\alpha)\right]$ such that

1. $\underline{a}(\alpha)$ is a bounded monotonic increasing left-continuous function $\forall \alpha \in (0, 1]$, right-continuous for $\alpha = 0$;
2. $\overline{a}(\alpha)$ is a bounded monotonic decreasing left-continuous function $\forall \alpha \in (0, 1]$, right-continuous for $\alpha = 0$;
3. $\underline{a}(\alpha) \le \overline{a}(\alpha), \forall \alpha \in [0, 1]$.

The notation $[A]_\alpha = \left[\underline{a}(\alpha), \overline{a}(\alpha)\right]$ is employed if such form exists.

2.2.1.1 Triangular Fuzzy Number

A triangular fuzzy number A has membership function (Fig. 2.2)

$$\mu_A = \begin{cases} \frac{x - x_l}{x_c - x_l}, & x \in (x_l, x_c] \\ \frac{x_r - x}{x_r - x_c}, & x \in (x_c, x_r) \\ 0, & otherwise \end{cases} \tag{2.9}$$

A triangular fuzzy number, which is generally identified by an ordered triplet of numbers (x_l, x_c, x_r), can be represented through the closed interval

$$[A]_\alpha = [(x_c - x_l)\alpha + x_l, (x_c - x_r)\alpha + x_r], \tag{2.10}$$

for any $\alpha \in [0, 1]$. A triangular fuzzy number is a reasonable mathematical model for the linguistic expression "nearly x_c" or "around x_c" if the triangular fuzzy number is symmetric.

In Scilab the function y=tri(1,c,r,x) returns the values of the membership degrees of a triangular fuzzy number, identified by the ordered triplet 1,c,r, once a vector of values x is provided.

Following (2.9), instructions for the left side, i.e.

```
x1 = x(1< x & x <= c);
y(find(1< x & x <= c)) = (x1 - 1)/(c-1);
```

and the right side

```
x1 = x(1< x & x <= c);
y(find(1< x & x <= c)) = (x1 - 1)/(c-1);
```

have to be provided.

The parametric form is quite easy to be implemented in Scilab. The function [yl,yr]=alpha_tri(1,c,r,a) returns the pair yl,yr, as given by the left-hand and right-hand formulas in (2.10), for any value of a representing α.

Refer to the Appendix for the complete listings.

Fig. 2.2 Triangular fuzzy number

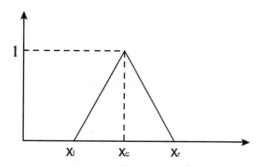

Example 2.4 The expression around five o'clock can be mathematically modeled by the symmetric triangular fuzzy number $A = (4.8, 5, 5.2)$. Its parametric representation is

$$[A]_\alpha = [0.2\alpha + 4.8, -0.2\alpha + 5.2],$$

Example 2.5 Let us consider the triangular fuzzy number $A = (1, 2, 3)$. Its parametric form is

$$[A]_\alpha = [\alpha + 1, -\alpha + 3],$$

For instance, for $\alpha = 0.5$, it is $[A]_{0.5} = [1, 5, 2.5]$, which is the corresponding α-cut.

Plotting in Scilab the α-cuts of $A = (1, 2, 3)$ for values of $\alpha \in \{0, 0.1, 0.2, \ldots, 0.9, 1\}$ (see Appendix), clearly returns the fuzzy triangular number A. In Fig. 2.3, a screenshot of the Scilab console is shown.

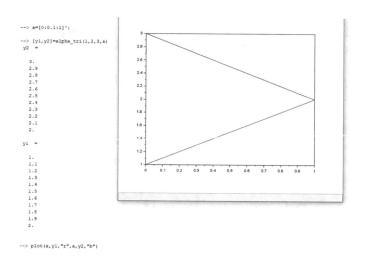

Fig. 2.3 α-levels of A in Scilab

The mentioned values of α are provided by the vector `a=[0:0.1:1]'`. Hence, the function `[y1,y2]=alpha_tri(1,2,3,a)` returns the computed pairs `y1,y2` for each entry of `a`. The line obtained by means of the `y1` values is plotted in red (`"r"` option in the plot command), while the line referred to the `y2` values is in blue (`"b"` option in the plot command).

2.2.1.2 Trapezoidal Fuzzy Number

A trapezoidal fuzzy number A, generally identified by (x_l, x_b, x_c, x_r), has membership function (Fig. 2.4)

$$\mu_A = \begin{cases} \frac{x-x_l}{x_b-x_l}, & x \in (x_l, x_b) \\ 1 & x \in [x_b, x_c] \\ \frac{x_r-x}{x_r-x_c}, & x \in (x_c, x_r) \\ 0, & otherwise \end{cases} \quad (2.11)$$

and it models the linguistic expression "approximately x_b to x_c" or "close to $[x_b, x_c]$"

The parametric form is

$$[A]_\alpha = [(x_b - x_l)\alpha + x_l, (x_c - x_r)\alpha + x_r] \quad (2.12)$$

for any $\alpha \in [0, 1]$.

The right trapezoid (2.13) models the linguistic expression "large" (Fig. 2.5)

$$\mu_A(x) = \begin{cases} 0 & if \ \ 0 \le x \le x_a, \\ \frac{x-x_a}{x_b-x_a} & if \ \ x_a < x < x_b, \\ 1 & if \ \ x_b \le x \le x_c. \end{cases} \quad (2.13)$$

Example 2.6 The fuzzy set of medium-sized flats can be represented by the trapezoidal fuzzy number $(60, 80, 90, 110)$ and its parametric form is

$$[A]_\alpha = [20\alpha + 60, -20\alpha + 110].$$

Fig. 2.4 Trapezoidal fuzzy number

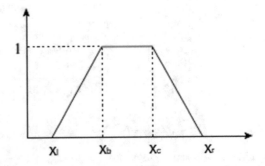

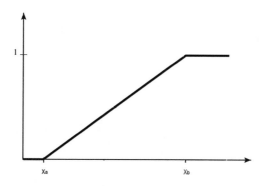

Fig. 2.5 Right-trapezoid shaped fuzzy number

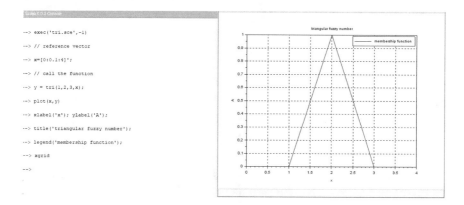

Fig. 2.6 Scilab plot: a triangular fuzzy number

In Scilab, the function `y=tra(l,b,c,r,x)` for a trapezoidal fuzzy number, identified by `l,b,c,r`, is very similar to the one for a triangular fuzzy number. The different part in it is referred to the core

```
x2=x(b<= x & x <=c);
y(find(b<= x & x <=c)) =1;
```

as one can check in the code listing in the Appendix.

Figures 2.6 and 2.7 show the screenshots of the console, where the plots of a triangular and a trapezoidal fuzzy number appear.

2.2.1.3 Bell-Shaped Fuzzy Number

A bell-shaped (Gaussian) fuzzy number has the following membership function

$$\mu_A = \begin{cases} \exp(-(\frac{x-x_c}{x_l})^2), & x \in [x_c - \delta, x_c + \delta] \\ 0, & otherwise \end{cases} \tag{2.14}$$

Fig. 2.7 Scilab plot: a trapezoidal fuzzy number

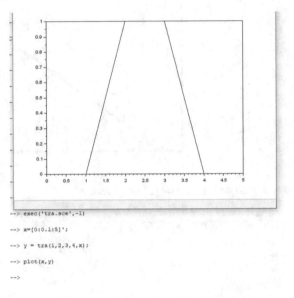

```
--> exec('tra.sce',-1)

--> x=[0:0.1:5]';

--> y = tra(1,2,3,4,x);

--> plot(x,y)

-->
```

Fig. 2.8 Fuzzy number with bell shape

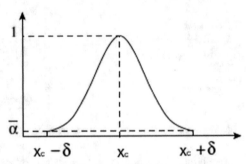

and models the linguistic expression "about x_c". The parametric form of fuzzy numbers in bell shape is given as follows:

$$[A]_\alpha = \begin{cases} \left[x_c - \sqrt{ln\left(\frac{1}{\alpha^{x_l^2}}\right)}, x_c + \sqrt{ln\left(\frac{1}{\alpha^{x_l^2}}\right)} \right], & if \quad \alpha \geq \overline{\alpha} \\ [x_c - \delta, x_c + \delta] \; if \quad \alpha < \overline{\alpha} \end{cases} \qquad (2.15)$$

where $\overline{\alpha} = \exp(-\frac{\delta}{x_l})^2$ and δ is the spread (Fig. 2.8).

2.2.1.4 LR-Type Fuzzy Number

A generalization of the triangular fuzzy numbers is offered by the LR-type fuzzy numbers (Fig. 2.9). Such fuzzy numbers were first introduced by Dubois and Prade (1978).

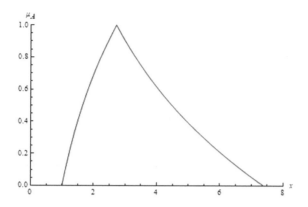

Fig. 2.9 An LR-type fuzzy number

A fuzzy number A is of LR-type if there exist functions L (for left), R (for right), and real parameters $\beta_1 > 0$, $\beta_2 > 0$ with

$$\mu_A(x) = \begin{cases} L\left(\frac{m-x}{\beta_1}\right) & if \ \ x \leq m \\ R\left(\frac{x-m}{\beta_2}\right) & if \ \ x > m \end{cases} \tag{2.16}$$

where the real parameter m is called the mean value of A, and β_1, β_2 are called the left and right spread, respectively.

Remark 2.2 A triangular fuzzy number can be regarded as LR-type with $m = x_c$, $\beta_1 = x_c - x_l$, $\beta_2 = x_r - x_c$ and

$$L(z) = 1 - z, \qquad R(z) = z - 1.$$

2.2.2 *Fuzzy Partition*

Let $I = [x_0, x_{m+1}]$ be a closed interval and $\{x_0, x_1, \ldots, x_{m+1}\}$, with $m \geq 3$, be points of I, called nodes, such that $x_0 < \cdots < x_{m+1}$. A fuzzy partition of I is a sequence $\{A_1, A_2, \ldots, A_m\}$ of continuous normal and convex fuzzy sets $A_i : I \to [0, 1]$, with $i = 1, \ldots, m$ such that $\sum_{i=1}^{m} A_i(x) = 1$, $\forall x \in I$. The fuzzy sets form a uniform fuzzy partition, when the nodes are equidistant. Figure 2.10 shows the uniform fuzzy partition obtained by triangular membership functions.

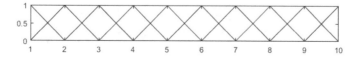

Fig. 2.10 Uniform fuzzy partition by triangular membership functions

Fuzzy partitions are used in fuzzy inference systems (Chap. 5), fuzzy neural networks and neuro-fuzzy systems (Chap. 6), fuzzy transforms (Chap. 7).

2.3 Interval Arithmetics

Let γ be a real number and, $A = [a_1, a_2]$ and $B = [b_1, b_2]$ two closed intervals on the real line.

Then the arithmetic operations between intervals can be defined as follows.
The *sum* of A and B is the interval

$$A + B = [a_1 + b_1, a_2 + b_2]. \tag{2.17}$$

The *difference* between A and B is the interval

$$A - B = [a_1 - b_2, a_2 - b_1]. \tag{2.18}$$

The *multiplication* of A by a scalar γ is the interval

$$\gamma A = \begin{cases} [\gamma a_1, \gamma a_2], & if \quad \gamma \geq 0, \\ [\gamma a_2, \gamma a_1], & if \quad \gamma < 0. \end{cases} \tag{2.19}$$

The *multiplication* of A by B is the interval

$$A \cdot B = [\min Q, \max Q], \tag{2.20}$$

where $Q = \{a_1 b_1, a_1 b_2, a_2 b_1, a_2 b_2\}$.
The *quotient* of A by B, with $B \subset \mathbb{R} - \{0\}$, is the interval

$$A/B = [a_1, a_2] \cdot \left[\frac{1}{b_2}, \frac{1}{b_1} \right]. \tag{2.21}$$

Example 2.7 Let $A = [-1, 2]$ and $B = [5, 6]$. Then:

$$A + B = [4, 8],$$

$$A - B = [-7, -3],$$

$$A \cdot B = [\min\{-5, -6, 10, 12\}, \max\{-5, -6, 10, 12\}] = [-6, 12],$$

$$A/B = \left[\min\left\{ -\frac{1}{6}, -\frac{1}{5}, \frac{1}{3}, \frac{2}{5} \right\}, \max\left\{ -\frac{1}{6}, -\frac{1}{5}, \frac{1}{3}, \frac{2}{5} \right\} \right] = \left[-\frac{1}{5}, \frac{2}{5} \right].$$

2.4 Arithmetic Operations with Fuzzy Numbers

2.4.1 First Method

A first method to perform arithmetic operations with fuzzy numbers is based on the Extension Principle.

Let A and B be two fuzzy numbers and γ a real number.

The *sum* of the fuzzy numbers A and B is the fuzzy number $A + B$, whose membership function is

$$\mu_{A+B}(z) = \sup_{\phi(z)} \min[\mu_A(x), \mu_B(y)], \qquad (2.22)$$

where $\phi(z) = \{(x, y) : x + y = z\}$.

The *difference* $A - B$ is the fuzzy number whose membership function is given by

$$\mu_{A-B}(z) = \sup_{\phi(z)} \min[\mu_A(x), \mu_B(y)], \qquad (2.23)$$

where $\phi(z) = \{(x, y) : x - y = z\}$.

The *multiplication* of A by B is the fuzzy number $A \cdot B$ whose membership function is given by

$$\mu_{A \cdot B}(z) = \sup_{\phi(z)} \min[\mu_A(x), \mu_B(y)], \qquad (2.24)$$

where $\phi(z) = \{(x, y) : xy = z\}$.

The *quotient* is the fuzzy number A/B whose membership function is

$$\mu_{A/B}(z) = \sup_{\phi(z)} \min[\mu_A(x), \mu_B(y)], \qquad (2.25)$$

where $\phi(z) = \{(x, y) : x/y = z\}$ and $0 \notin supp(B)$.

The *multiplication* of A by a scalar $\gamma \neq 0$ is the fuzzy number γA whose membership function is given by

$$\mu_{\gamma A}(z) = \begin{cases} \sup_{\phi(z)}[\mu_A(x)], & if \quad \gamma \neq 0 \\ \chi_0(z), & if \quad \gamma = 0 \end{cases} = \begin{cases} \mu_A(\gamma^{-1}z), & if \quad \gamma \neq 0 \\ \chi_0(z), & if \quad \gamma = 0 \end{cases} \qquad (2.26)$$

where $\phi(z) = \{x : \gamma x = z\}$ and $\chi_0(z)$ is the characteristic function of $\{0\}$.

Remark 2.3 For any commutative operation on real numbers, the extended operation on fuzzy numbers is also commutative.

For any associative operation on real numbers, the extended operation on fuzzy numbers is also associative.

Example 2.8 Consider the finite fuzzy sets

$$A = \left\{ \frac{0.5}{1} + \frac{1}{2} + \frac{0.7}{3} \right\},$$

$$B = \left\{ \frac{0.6}{4} + \frac{1}{6} + \frac{0.7}{8} \right\},$$

Perform the division A/B by using the Extension Principle.

$$A/B = \left\{ \frac{0.5}{1} + \frac{1}{2} + \frac{0.6}{3} \right\} / \left\{ \frac{0.6}{4} + \frac{1}{6} + \frac{0.7}{8} \right\} =$$

$$= \max \left\{ \frac{\min(0.5, 0.6)}{1/4} + \frac{\min(0.5, 1)}{1/6} + \cdots + \frac{\min(1, 0.6)}{2/4} + \cdots + \right.$$

$$\left. + \frac{\min(1, 0.7)}{2/8} + \cdots + \frac{\min(0.7, 1)}{3/6} + \frac{\min(0.7, 0.7)}{3/8} \right\} =$$

$$= \max \left\{ \frac{0.5}{1/4} + \frac{0.5}{1/6} + \cdots + \frac{0.6}{1/2} + \cdots + \frac{0.7}{1/4} + \cdots + \frac{0.7}{1/2} + \frac{0.7}{3/8} \right\} =$$

$$= \left\{ \frac{0.5}{1/6} + \frac{0.5}{1/8} + \frac{1}{1/3} + \frac{0.7}{1/4} + \frac{0.6}{3/4} + \frac{0.7}{1/2} + \frac{0.7}{3/8} \right\}.$$

2.4.2 Second Method

A practical method for operations with fuzzy numbers is based on interval arithmetic and α-cuts.

It is possible to prove (Pedrycz and Gomide 1998) that the α-levels of the fuzzy set $A \otimes B$ are given by

$$[A \otimes B]_\alpha = [A]_\alpha \otimes [B]_\alpha \tag{2.27}$$

$\forall \alpha \in [0, 1]$ and any arithmetic operation $\otimes \in \{+, -, \ldots, /\}$.

Proposition 2.1 Let A and B be fuzzy numbers with their parametric representation $[A]_\alpha = [\underline{a}(\alpha), \overline{a}(\alpha)]$ and $[B]_\alpha = [\underline{b}(\alpha), \overline{b}(\alpha)]$. Then the following properties hold.

(a) The sum of A and B is the fuzzy number $A + B$ with parametric form

$$[A + B]_\alpha = [\underline{a}(\alpha) + \underline{b}(\alpha), \overline{a}(\alpha) + \overline{b}(\alpha)]. \tag{2.28}$$

(b) The difference of A and B is the fuzzy number $A - B$ with parametric form

$$[A - B]_\alpha = \left[\underline{a}(\alpha) - \overline{b}(\alpha), \overline{a}(\alpha) - \underline{b}(\alpha)\right]. \tag{2.29}$$

(c) The multiplication of A by a scalar γ is the fuzzy number γA with parametric form

$$[\gamma A]_\alpha = \gamma [A]_\alpha = \begin{cases} \left[\gamma\underline{a}(\alpha), \gamma\overline{a}(\alpha)\right], & if \quad \gamma \geq 0 \\ \left[\gamma\overline{a}(\alpha), \gamma\underline{a}(\alpha)\right], & if \quad \gamma < 0. \end{cases} \tag{2.30}$$

(d) The multiplication of A by B is the fuzzy number $A \cdot B$ with parametric form

$$[A \cdot B]_\alpha = [A]_\alpha [B]_\alpha = [\min P_\alpha, \max P_\alpha], \tag{2.31}$$

where $P_\alpha = \{\underline{a}(\alpha)\underline{b}(\alpha), \underline{a}(\alpha)\overline{b}(\alpha), \overline{a}(\alpha)\underline{b}(\alpha), \overline{a}(\alpha)\overline{b}(\alpha)\}$.

(e) The division of A by B, with $0 \notin supp(B)$, is the fuzzy number $A \cdot B$ with parametric form

$$\left[\frac{A}{B}\right]_\alpha = \frac{[A]_\alpha}{[B]_\alpha} = \left[\underline{a}(\alpha), \overline{a}(\alpha)\right]\left[\frac{1}{\overline{b}(\alpha)}, \frac{1}{\underline{b}(\alpha)}\right]. \tag{2.32}$$

Notice that with triangular fuzzy numbers, the sum, the difference and the multiplication by a scalar result in a triangular fuzzy number. Let $A = (a_1, a_c, a_2)$ and $B = (b_1, b_c, b_2)$ be two fuzzy numbers, with their parametric form

$$[A]_\alpha = [(a_c - a_1)\alpha + a_1, (a_c - a_2)\alpha + a_2], \quad [B]_\alpha = [(b_c - b_1)\alpha + b_1, (b_c - b_2)\alpha + b_2]. \tag{2.33}$$

Then

$$[A + B]_\alpha = [(a_c + b_c - a_1 - b_1)\alpha + (a_1 + b_1), (a_c + b_c - a_2 - b_2)\alpha + (a_2 + b_2)], \tag{2.34}$$

and the resulting triangular fuzzy number is $A + B = (a_1 + b_1, a_c + b_c, a_2 + b_2)$. Also notice that $(A - B) + B \neq A$.

Performing the addition and the subtraction of two triangular fuzzy numbers in Scilab is quite easy (see Example 2.9). Instead, the multiplication and, especially the division, require more attention.

The function $[yl, yr] = alpha\_tri\_m(l1, c1, r1, l2, c2, r2, a)$ returns the pairs to obtain the left-hand and right-hand side of the fuzzy set resulting from the multiplication of the triangular fuzzy numbers $(l1, c1, r1)$ and $(l2, c2, r2)$, given the entries of the vector a. Once the alpha levels of the two triangles A and B, identified by the above-mentioned triplets, have been computed as follows

```
A=[(c1-l1)*a+l1,(c1-r1)*a+r1];
B=[(c2-l2)*a+l2,(c2-r2)*a+r2];
```

the set of products P needs to be formed

```
for i=1:2
for j=1:2
P(:,k)=A(:,j).*B(:,i);
k=k+1;
end
end
```

and eventually sorting this set, it is possible to get the extremal values `y1` and `yr` (see the complete code in the Appendix). The function for the division of two triangular fuzzy sets `[y1,yr]=alpha_tri_d(l1,c1,r1,l2,c2,r2,a)` is similar, but with different entries for P:

```
P(:,k)=A(:,j)./B(:,i);
```

and an initial control to check and exclude the entry of the vector a which makes a denominator in P null

```
if r2==0 |l2==0 then
a(find(a==0))=[];
else if l2<0 & r2>0 then
a(find(a==-l2/(c2-l2)))=[];
end
end
```

Example 2.9 Let $A = (-1, 1, 4)$ and $B = (2, 4, 6)$ be two triangular fuzzy number, with parametric form

$$[A]_\alpha = [2\alpha - 1, 4 - 3\alpha], \qquad [B]_\alpha = [2\alpha + 2, 6 - 2\alpha].$$

Then

- $[A + B]_\alpha = [4\alpha + 1, 10 - 5\alpha], \qquad A + B = (1, 5, 10);$
- $[A - B]_\alpha = [-3, -2 - \alpha], \qquad A - B = (-3, -3, -2).$

In Scilab, one has to write the following instructions to get the sum:

```
a=[0:0.1:1]';
exec('alpha_tri.sce');
[ya1,ya2]=alpha_tri(-1,1,4,a);
[yb1,yb2]=alpha_tri(2,4,6,a);
S=[ya1,ya2]+[yb1,yb2]
plot(a,S(:,1),"r",a,S(:,2),"b")
```

while for the difference, it is

```
D=[ya1,ya2]-[yb1,yb2]
plot(a,D(:,1),"r",a,D(:,2),"b")
```

The results and the plots are shown in Figs. 2.11 and 2.12.

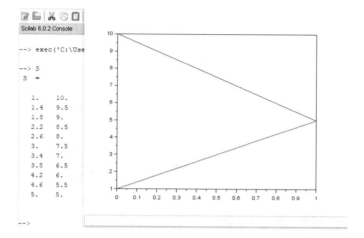

Fig. 2.11 Sum of two fuzzy triangular numbers

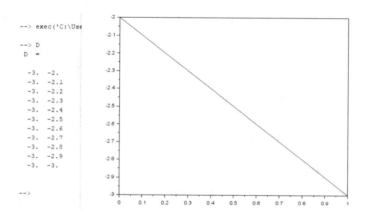

Fig. 2.12 Difference of two fuzzy triangular numbers

Example 2.10 Let $A = (1, 2, 3)$ and $B = (3, 4, 5)$ be two triangular fuzzy numbers. In parametric form:

$$[A]_\alpha = [1 + \alpha, 3 - \alpha], \quad [B]_\alpha = [3 + \alpha, 5 - \alpha].$$

The multiplication is

$$[A \cdot B]_\alpha = [(1 + \alpha)(3 + \alpha), (3 - \alpha)(5 - \alpha)].$$

This multiplication can be performed in Scilab. The screenshot showing the outcome is depicted in Fig. 2.13.

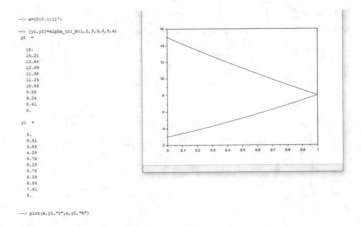

```
--> a=[0:0.1:1]';

--> [y1,y2]=alpha_tri_m(1,2,3,3,4,5,a)
 y2  =

  15.
  14.21
  13.44
  12.69
  11.96
  11.25
  10.56
  9.89
  9.24
  8.61
  8.

 y1  =

  3.
  3.41
  3.84
  4.29
  4.76
  5.25
  5.76
  6.29
  6.84
  7.41
  8.

--> plot(a,y1,"r",a,y2,"b")
```

Fig. 2.13 Multiplication of two fuzzy triangular numbers

Example 2.11 Let $A = (9, 10, 11)$ and $B = (3, 10, 12)$ be two triangular numbers. The parametric form is:

$$[A]_\alpha = [\alpha + 9, \alpha + 11], \ [B]_\alpha = [7\alpha + 3, 2\alpha + 12].$$

Let us consider the following operation:

$$(10A)/(10B) = \left[\frac{10\alpha + 90}{120 - 20\alpha}, \frac{110 - 10\alpha}{70\alpha + 30} \right].$$

Figure 2.14 shows the result of the division above in Scilab.

Notice that when excluding zero from the divisor's support, it is possible to get a result but a kind of discontinuity appears (see Fig. 2.15).

2.4.3 Method for LR-Type Fuzzy Numbers

It is possible to prove the following (Dubois and Prade 1980).

Proposition 2.2 Let $A = (m, \beta_1, \beta_2)_{LR}$ and $B = (n, \delta_1, \delta_2)_{LR}$ be two LR-type fuzzy numbers. Then

$$A + B = (m + n, \beta_1 + \delta_1, \beta_2 + \delta_2)_{LR}, \tag{2.35}$$

$$-A = (-m, \beta_2, \beta_1)_{LR}, \tag{2.36}$$

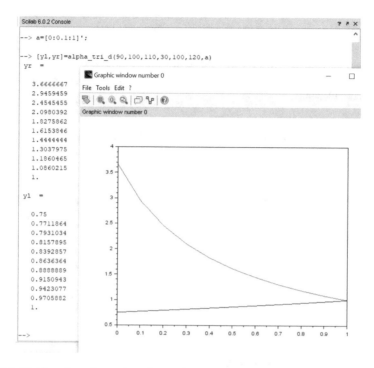

Fig. 2.14 Division of two fuzzy triangular numbers: a particular case

Fig. 2.15 Division of two
fuzzy triangular numbers

```
a=[0:0.01:1]';
[y1,y2]=alpha_tri_d(1,2,3,-1,1,3,a)
```

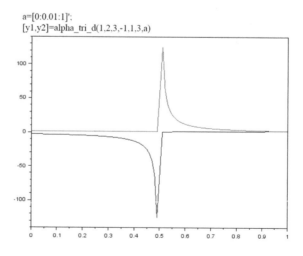

$$A - B = (m - n, \beta_1 + \delta_2, \beta_2 + \delta_1)_{LR}. \tag{2.37}$$

Proposition 2.3 Let $A = (m, \beta_1, \beta_2)_{LR}$ and $B = (n, \delta_1, \delta_2)_{LR}$ be two LR-type fuzzy numbers. Then

-
$$A \cdot B = (mn, n\beta_1 + m\delta_1, n\beta_2 + m\delta_2)_{LR}, \tag{2.38}$$

for A, B positive;

-
$$A \cdot B = (mn, n\beta_1 - m\delta_2, n\beta_2 - m\delta_1)_{LR}, \tag{2.39}$$

for A negative, B positive;

-
$$A \cdot B = (mn, -n\beta_2 - m\delta_2, n\beta_1 - m\delta_1)_{LR}, \tag{2.40}$$

for A, B negative.

Problems

1. Plot the complement of a Gaussian membership function such that it is not null in any point.
2. Plot the LR membership function, with $L(x) = \max(0, 1 - x)$ and $R(x) = \exp(-x^2)$ for different values of the parameters.
3. Implement in Scilab the complement of the fuzzy set described by the following S-shaped membership function

$$\mu_{\gamma A}(z) = \begin{cases} 0, & if \ x \le \lambda, \\ 2(\frac{x-\lambda}{\rho-\lambda})^2, & if \ \lambda < x \le \frac{\lambda+\rho}{2}, \\ 1 - 2(\frac{\rho-x}{\rho-\lambda})^2, & if \ \frac{\lambda+\rho}{2} < x \le \rho, \\ 1, & if \ \rho < x. \end{cases}$$

Try different values of the parameters λ and ρ.
4. Consider the finite fuzzy sets

$$A = \left\{ \frac{0.5}{1} + \frac{1}{2} + \frac{0.7}{3} \right\},$$

$$B = \left\{ \frac{0.6}{4} + \frac{1}{6} + \frac{0.7}{8} \right\}.$$

Perform the addition $A + B$ and the product $A \cdot B$ by using the Extension Principle.

References

Dubois D, Prade H (1978) Operations on fuzzy numbers. Int J Syst Sci 9:613–626

Dubois D, Prade H (1980) Fuzzy sets and systems: theory and applications, Academic Press, New York

Pedrycz W, Gomide F (1998) An Introduction to Fuzzy Sets: Analysis and Design, The MIT Press, Massachusets

Zadeh LA (1975) The concept of a linguistic variable and its application to approximate reasoning I, II, III. Inf Sci 8:199–249, 301–357, 43–80

Chapter 3
Fuzzy Relations

3.1 Fuzzy Relations on Sets and Fuzzy Sets

3.1.1 Relations and Fuzzy Relations

The concept of relation in mathematics belongs to the set theory. A classical relation indicates whether there is or not any association between some elements, while fuzzy relations indicate, in addition, the degree of this association. The choice of the relation depends on the considered phenomenon. Fuzzy relations can be regarded as useful tools in problems of information retrieval, pattern classification, control and decision-making.

A relation R on $X_1 \times \cdots \times X_r$ is any subset of the Cartesian product of the sets $X_1, \ldots, X_r$. If $r = 2$, then the relation is called a binary relation on $X_1 \times X_2$. If $X_1 = X_2 = \cdots = X_n = X$, then R is a n-ary relation on X.

A classical relation R can be represented by its characteristic function

$$\chi_R : X_1 \times \ldots \times X_r \to \{0, 1\}, \tag{3.1}$$

with

$$\chi_R(x_1, x_2, \ldots, x_r) = \begin{cases} 1, & if \quad (x_1, x_2, \ldots, x_r) \in R \\ 0, & if \quad (x_1, x_2, \ldots, x_r) \notin R. \end{cases} \tag{3.2}$$

Definition 3.1 A fuzzy relation R on $X_1 \times \cdots \times X_r$ is any fuzzy subset of $X_1 \times \cdots \times X_r$. A fuzzy relation R is defined by a membership function

$$\mu_R : X_1 \times \cdots \times X_r \to [0, 1] \tag{3.3}$$

If there are only two sets $X_1 \times X_2$, the relation is called a binary fuzzy relation.

Let R be a fuzzy relation on $X \times Y$. If $R(x, y) = 1$, then the two elements $x \in X$ and $y \in Y$ are fully related. $R(x, y) = 0$ means that these elements are unrelated while the values in-between 0 and 1 underline a partial association between x and y.

© The Author(s), under exclusive license to Springer Nature Switzerland AG 2022
S. Tomasiello et al., *Contemporary Fuzzy Logic*, Big and Integrated
Artificial Intelligence 1, https://doi.org/10.1007/978-3-030-98974-3_3

Similarly as in the case of fuzzy sets, fuzzy relations can be represented by their α-cuts.

When X and Y are finite sets with cardinality $|X| = p$ and $|Y| = q$, then a fuzzy relation can be easily represented as follows. Let $X = \{x_1, x_2, \ldots, x_p\}$ and $Y = \{y_1, y_2, \ldots, y_q\}$ be two finite sets, then a relation R in $X \times Y$ can be arranged in tabular or matrix form, with $p \times q$ entries $r_{ij} \in [0, 1]$:

$$
\begin{array}{c|ccc}
R & y_1 & \cdots & y_q \\
\hline
x_1 & r_{11} & \cdots & r_{1q} \\
\vdots & \vdots & \vdots & \vdots \\
x_p & r_{m1} & \cdots & r_{pq}
\end{array}
$$

$$
R = \begin{bmatrix} r_{11} & \cdots & r_{1q} \\ \vdots & \vdots & \vdots \\ r_{m1} & \cdots & r_{pq} \end{bmatrix} \tag{3.4}
$$

Each entry r_{ij} represents the corresponding degree of association between x_i and y_j.

Definition 3.2 Let R be a binary fuzzy relation defined on $X \times Y$. The inverse binary fuzzy relation, R^{-1}, defined on $Y \times X$, has the following membership function $\mu_{R^{-1}} : Y \times X \to [0, 1]$, with $\mu_{R^{-1}}(y, x) = \mu_R(x, y)$.

Notice that the matrix of R^{-1} coincides with the transpose of R.

Let R be a (classical) binary relation over U, with characteristic function χ_R. Then, for any x, y and z of U, the relation R is

(i) reflexive, if $\chi_R(x, x) = 1$;
(ii) symmetric, if $\chi_R(x, y) = 1$ implies $\chi_R(y, x) = 1$;
(iii) transitive, if $\chi_R(x, y) = \chi_R(y, z) = 1$ implies $\chi_R(x, z) = 1$;
(iv) anti-symmetric, if $\chi_R(x, y) = \chi_R(y, x) = 1$ implies $x = y$.

Let R be a binary fuzzy relation over U, with membership function μ_R. Then, for any x, y and z of U, the fuzzy relation R is

(i) reflexive, if $\mu_R(x, x) = 1$;
(ii) symmetric, if $\mu_R(x, y) = \mu_R(y, x)$;
(iii) transitive, if $\mu_R(x, z) \geq \mu_R(x, y) \wedge \mu_R(y, z)$, where $\wedge$ is the minimum;
(iv) anti-symmetric, if $\mu_R(x, y) > 0$ and $\mu_R(y, x) > 0$ implies $x = y$.

For example, the relation $R :=$ "is relative of" is reflexive, symmetric but not transitive.

Definition 3.3 A similarity relation is a fuzzy relation μ_s that is reflexive, symmetric and transitive.

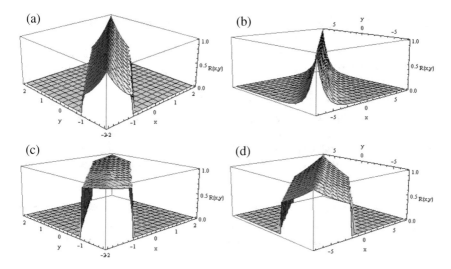

Fig. 3.1 Example 3.1: **a** $\beta = 1, \delta = 1$ **b** $\beta = 1, \delta = 5$ **c** $\beta = 10, \delta = 1$ **d** $\beta = 10, \delta = 5$

Example 3.1 $R :=$ "approximately equal" (or similar)

$$\mu_R(x, y) = \begin{cases} \exp(-|x - y|/\beta), & \text{for } |x - y| \leq \delta \\ 0, & \text{otherwise} \end{cases}$$

for $\beta, \delta > 0$. In Fig. 3.1, (b) is particularly suitable to express the relation "approximately equal".

Example 3.2 With regard to Example 3.1, considering the finite sets $X = \{2, 4\}$ and $Y = \{4, 7, 10\}$, with $\beta = 10$, the discrete form of the previous relation μ_R is

R	y_1	y_2	y_3
x_1	0.818731	0.606531	0.449329
x_2	1	0.740818	0.548812

It can also be written as a matrix:

$$R = \begin{bmatrix} 0.818731 & 0.606531 & 0.449329 \\ 1 & 0.740818 & 0.548812 \end{bmatrix}.$$

The inverse R^{-1} is the transpose of R:

$$R^{-1} = \begin{bmatrix} 0.818731 & 1 \\ 0.606531 & 0.740818 \\ 0.449329 & 0.548812 \end{bmatrix}.$$

Example 3.3 In this example, we consider a fuzzy relation for "resemblance"

$$\mu_R(x, y) = \begin{cases} \exp(-((x-1)^2 + (y-1)^2)/10), & \text{for } (x, y) \in X \times Y \\ 0, & \text{otherwise} \end{cases}$$

with $X = [-10, 10]$ and $Y = [-10, 10]$ (Fig. 3.2).

Let us consider the finite sets $X = \{1, 2, 3\}$ and $Y = \{2.25, 3.75\}$, then the discrete relation for resemblance can be expressed in tabular form. In particular, by replacing the numerical values of the elements of X and Y with names, one gets Table 3.1.

This discrete fuzzy relation can also be represented by a diagram as shown in Fig. 3.3.

Example 3.4 Let i_1, i_2, i_3, i_4 be four items. Users can have certain preferences, let us say w_1, w_2, w_3, considering 3 users. Then a relation R on $I \times W$, with $I = \{i_1, i_2, i_3, i_4\}$ and $W = \{w_1, w_2, w_3\}$ can assume the matrix form with the following entries

$$R = \begin{bmatrix} 1 & 0 & 0.6 \\ 0.8 & 1 & 0 \\ 0 & 1 & 0 \\ 0.8 & 0 & 1 \end{bmatrix}.$$

Fig. 3.2 A fuzzy relation for "resemblance"

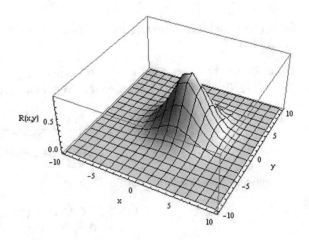

Table 3.1 A resemblance relation

	Paul	George
Luke	0.855	0.469
Sam	0.774	0.425
Andrew	0.573	0.315

Fig. 3.3 Fuzzy relation as a diagram

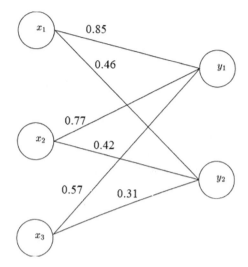

3.1.2 Fuzzy Relations on Fuzzy Sets

Definition 3.4 Let $X, Y \subseteq \mathbb{R}$ and

$$A = \{(x, \mu_A(x)|x \in X\}, \qquad B = \{(y, \mu_B(y)|y \in Y\}, \qquad (3.5)$$

be two fuzzy sets. Then $R = \{[(x, y), \mu_R(x, y)]|(x, y) \in X \times Y\}$ is a fuzzy relation on A and B if

$$\mu_R(x, y) \leq \mu_A(x), \forall (x, y) \in X \times Y \qquad (3.6)$$

and

$$\mu_R(x, y) \leq \mu_B(y), \forall (x, y) \in X \times Y. \qquad (3.7)$$

Definition 3.5 The fuzzy Cartesian product of the fuzzy sets $A_1, A_2, \ldots, A_r$ of $X_1, \ldots, X_r$, respectively, is the fuzzy relation $A_1 \times A_2 \times \cdots \times A_r$, whose membership function is

$$\mu_{A_1 \times A_2 \times \ldots \times A_r}(x_1, x_2, \ldots, x_r) = \mu_{A_1}(x_1) \wedge \mu_{A_2}(x_2) \ldots \wedge \mu_{A_r}(x_r), \qquad (3.8)$$

where $\wedge$ represents the minimum.

3.2 Composition of Fuzzy Relations

Fuzzy relations in different product spaces can be combined by the operation *composition*. There are different types of composition. The max-min composition is the most frequently used one.

Definition 3.6 Let $R_1(x, y)$, $(x, y) \in X \times Y$ and $R_2(y, z)$, $(y, z) \in Y \times Z$ be two fuzzy relations. The *max-min composition* $R_1 \quad max - min \quad R_2$ is a binary fuzzy relation in $X \times Z$ whose membership function is given by

$$\mu_{R_1 \circ R_2}(x, z) = \max_{y}\{\min\{\mu_{R_1}(x, y), \mu_{R_2}(y, z)\}\} \tag{3.9}$$

Let the sets X, Y and Z be finite. The matrix form of the relation given by the max-min composition can be regarded as a matrix multiplication, replacing the product by the minimum and the sum by the maximum. Then, given two fuzzy relations R and S, the max-min composition matrix $T = R \circ S$ has entries

$$t_{ij} = \max_{k}[\min(r_{ik}, s_{kj})]. \tag{3.10}$$

The Scilab function C = max_min(A,B) implements (3.10) as follows

```
for i = 1:m
for j = 1:q
tm= min(A(i, :), B(:, j)');
C(i,j) = max(tm);
end
end
```

where m is the number of rows of A and q the number of columns of B. When the number of columns of A and the number of rows of B are different, an error message about the incompatible sizes is returned.

Example 3.5 Let A and B be two fuzzy relations, given by the two following matrices:

$$A = \begin{bmatrix} 0.1 \ 0.2 \ 0.3 \\ 0.4 \ 0.5 \ 0.6 \end{bmatrix},$$

$$B = \begin{bmatrix} 0.7 \ 0.8 \ 0.9 \\ 0.1 \ 0.2 \ 0.3 \\ 0.4 \ 0.5 \ 0.6 \end{bmatrix}.$$

The composition $A \ max - min \ B$ gives the matrix C

$$C = \begin{bmatrix} 0.3 \ 0.3 \ 0.3 \\ 0.4 \ 0.5 \ 0.6 \end{bmatrix}.$$

By performing this max-min composition by Scilab, one gets the result shown in Fig. 3.4.

Fig. 3.4 Max-min
composition in Scilab

```
--> A=[0.1 0.2 0.3; 0.4 0.5 0.6];

--> B=[0.7 0.8 0.9; 0.1 0.2 0.3; 0.4 0.5 0.6];

--> C=max_min(A,B)
 C  =

    0.3    0.3    0.3
    0.4    0.5    0.6


-->
```

3.2.1 Set-Relation Composition

Set-relation compositions are particularly important in fuzzy inference systems. The max-min composition of a fuzzy set A over X and a fuzzy relation R over $X \times Y$ is a fuzzy set B over Y, $B = A \circ R$, defined by the following membership function

$$\mu_B(y) = \max_{x \in X} min[\mu_A(x), \mu_R(x, y)], \qquad \forall y \in Y. \tag{3.11}$$

Example 3.6 Let $X = Y = \{10, 8, 6, 4\}$ be the universe. Let A be a finite fuzzy set representing the term *large*:

$$A = large = \{(10, 1), (8, 0.6), (6, 0.2), (4, 0)\}.$$

Let R be the relation *approximately equal*, expressed as follows

$$R = \begin{bmatrix} 1 & 0.5 & 0 & 0 \\ 0.5 & 1 & 0.5 & 0 \\ 0 & 0.5 & 1 & 0.5 \\ 0 & 0 & 0.5 & 1 \end{bmatrix}.$$

Consider the set-relation composition $B = A \circ R$. This operation gives

$$B = \{(10, 1), (8, 0.6), (6, 0.5), (4, 0.2)\},$$

where the fuzzy set B represents *approximately large*.

By using Scilab to solve this exercise, the result shown in Fig. 3.5 is obtained.

Fig. 3.5 Set-relation
composition in Scilab

```
--> A=[1 0.6 0.2 0];

-->

--> R=[1 0.5 0 0;0.5 1 .5 0; 0 0.5 1 0.5;0 0 0.5 1];

-->

--> B=max_min(A,R)
 B  =

    1.    0.6    0.5    0.2
```

3.3　Fuzzy Relational Equations

Let us consider the finite universes $U = \{u_1, u_2, \ldots, u_m\}$, $V = \{v_1, v_2, \ldots, v_n\}$ and $W = \{w_1, w_2, \ldots, w_p\}$. We consider fuzzy relational equations of the form

$$R * X = T, \quad \text{or} \quad X * R = T \tag{3.12}$$

where R, T, X denote the matrix form of given binary fuzzy relations in $U \times V$, $U \times W$, $V \times W$ respectively, "$*$" stands for any fuzzy relational composition. When R and T are known while X is the unknown fuzzy relation, an *inverse problem* has to be solved. In the *estimation problem*, given X and T, one has to determine the fuzzy relation R. For example, X could be the collection of the possible causes for abnormal patterns in a manufacturing process and T the collection of the possible abnormal patterns, and one wants to find the relationship between abnormal patterns and their causes.

3.3.1　Fuzzy Relational Equations with the Max-Min Composition

Let us consider the equation

$$R \circ X = T, \tag{3.13}$$

where $\circ$ denotes the max-min composition. Assuming that the universes are finite, the fuzzy relations have matrix representation $R = [r_{ij}]$, $X = [x_{jk}]$ and $T = [t_{ik}]$. where $r_{ij} = \mu_R(u_i, v_j)$, $x_{jk} = \mu_X(v_j, w_k)$, $t_{ik} = \mu_T(u_i, w_k)$. Then the problem is to find $x_{jk} \in [0, 1]$ such that

$$\max_{1 \leq j \leq n} [\min(r_{ij}, x_{jk})] = t_{ik}, \tag{3.14}$$

for any $1 \leq i \leq m$ and $1 \leq k \leq p$.

Proposition 3.1 (de Barros 2017) If the following inequality

$$\max_{1 \leq j \leq n} r_{ij} < \max_{1 \leq k \leq p} t_{ik} \tag{3.15}$$

holds true, then the fuzzy relational equation (3.13) has no solution.

Remark 3.1 For equations in the form $Y \circ R = T$, where Y is the unknown fuzzy relation, the solution is the inverse (transpose of the resulting matrix) of the solution of $R^{-1} \circ X = T^{-1}$.

A maximal solution of the fuzzy relational equations is a fuzzy relation that solves the given equation where each element has the highest membership degree. Let us consider the fuzzy relational equation

$$x \circ R = t. \tag{3.16}$$

for an $m \times n$ matrix R and n-dimensional vector t.

The maximal solution of the fuzzy relational equation (3.16) is $\overline{x} = (\overline{x}_1, \ldots, \overline{x}_m)$, where

$$\overline{x}_i = \begin{cases} t_j, & \text{if } r_{ij} > t_j \\ 1, & \text{otherwise} \end{cases} \tag{3.17}$$

for any $j \in \{1, \ldots, n\}$, with $i \in \{1, \ldots, m\}$ (Higashi and Klir 1984). Notice that by taking $x_h = (x_{h1}, x_{h2}, \ldots, x_{hm})$ and $t_h = (t_{h1}, t_{h2}, \ldots, t_{hn})$, then the fuzzy relational equation $X \circ R = T$ can be reduced to the set of equations $x_h \circ R = t_h$.

Example 3.7 Let us consider the following fuzzy relational equation:

$$[x_1 \quad x_2] \circ \begin{bmatrix} 0.7 & 0.2 \\ 0.3 & 0.6 \end{bmatrix} = [0.5 \quad 0.5].$$

As one can see, $R_{11} = 0.7$ and $R_{22} = 0.6$ are both greater than $t_1 = t_2 = 0.5$, then according to (3.17) the maximal solution is $[0.5 \quad 0.5]$ and this can be easily checked:

$$\max(\min(0.7, x_1), \min(0.3, x_2)) = 0.5$$

$$\max(\min(0.2, x_1), \min(0.6, x_2)) = 0.5$$

3.3.2 Bilinear Fuzzy Relation Equations

Bilinear fuzzy relation equations have the following form:

$$A \circ X = B \circ X, \tag{3.18}$$

where A and B are fuzzy relations, expressed as $m \times n$ matrices, and X is unknown. Let $\circ$ denote the max-min composition and $\overline{a}_i, \overline{b}_i \in [0, 1]$ be two assigned real numbers. The system

$$\overline{a}_i \vee \bigcup_{j=1}^{n} (a_{ij} \wedge x_j) = \overline{b}_i \vee \bigcup_{j=1}^{n} (b_{ij}) \tag{3.19}$$

is called *system of external fuzzy bilinear equations*. If $\overline{a}_i = \overline{b}_i = 0$, then the system (3.19) reduces to the *system of fuzzy bilinear equations*

$$\bigcup_{j=1}^{n} a_{ij} \wedge x_j = \bigcup_{j=1}^{n} b_{ij} \wedge x_j. \tag{3.20}$$

The smallest solution of this system is the trivial solution.

3.3.2.1 Bilinear Fuzzy Relation Equations

The following ρ_i are called critical values of the ith fuzzy bilinear equation

$$\rho_i = \min \left(\bar{a}_i \vee \bigcup_{j=1}^{n} a_{ij}, \bar{b}_i \vee \bigcup_{j=1}^{n} b_{ij} \right). \tag{3.21}$$

Let $I_n = \{1, 2, \ldots, n\}$. The sets

$$\Delta_i^1 = \{j \in I_n : b_{ij} > \rho_i\}, \qquad \Delta_i^2 = \{j \in I_n : a_{ij} > \rho_i\} \tag{3.22}$$

are called the difference index sets of critical value of the ith fuzzy bilinear equation (3.19).

The following algorithm for the solution of bilinear fuzzy relation equations has been proposed by Di Martino and Sessa (2018).

1. Compute $\rho_k = \min\{\rho_i : i = 1, \ldots, m\}$ and the corresponding set $\Delta_k = \Delta_k^1 = \{j_1, j_2, \ldots, j_t\}$.
2. Put $\bar{x}_{j_h} = \rho_k$ for $h = 1, \ldots, t$, (implying to get a system of $m - t$ external fuzzy bilinear equations with $m - t$ variables obtained by replacing the variables $x_{j_1}, x_{j_2}, \ldots, x_{j_t}$ with ρ_k). If $\Delta_k = I_n$, it is $\bar{x} = (\rho_k, \ldots, \rho_k)$ otherwise store the variable $\Delta = I_n - \Delta_k$.
3. Consider the new system given by $m - t$ external fuzzy bilinear equations and calculate the values ρ_i, ρ_k.
4. For $\Delta_k = \{j_1, j_2, \ldots, j_s\}$, put $\bar{x}_{j_h} = \rho_k$ for $h = 1, \ldots, s$ and update the variable Δ as $\Delta = \Delta - \Delta_k$. Obtain a new system of $\bar{m} = m - t - s$ external bilinear equations with $\bar{m}$ variables. If $\bar{m} > 0$, go back to the step 3, otherwise set $\bar{x}_h = 1$ for any $h \in \Delta$ and stop.

Problems

1. Show an example fuzzy relation describing linkages among

    (i) restaurant-goers $\{a, b, c, d\}$ and restaurants $\{r1, r2, \ldots, r7\}$
    (ii) teenagers $\{a, b, c, d\}$, books $\{b1, b2, \ldots, b10\}$, and locations of the bookstores $\{A, B, C, D, E\}$.

2. Show an example of a set-relation composition with a finite fuzzy set representing *medium* and a fuzzy relation *much smaller than*. Elaborate on the outcome.

3. Find the maximal solution (if it exists) of the fuzzy relational equation

$$[x_1 \quad x_2 \quad x_3] \circ \begin{bmatrix} 1 & 0.2 & 0.4 \\ 0.3 & 1 & 0.7 \\ 0.6 & 0.5 & 0.8 \end{bmatrix} = [0.5 \quad 0.5 \quad 0.7].$$

4. Write a Scilab code to check whether a fuzzy relational equation in the form (3.13) has no solution.

References

de Barros LC, Bassanezi RC, Lodwick WA (2017) A first course in fuzzy logic, fuzzy dynamical systems, and biomathematics, theory and applications. Springer, Heidelberg

Di Martino F, Sessa S (2018) Comparison between images via bilinear fuzzy relation equations. J Ambient Intell Hum Comput 9:1517–1525

Higashi M, Klir G (1984) Resolution of finite fuzzy relation equations. Fuzzy Sets Syst 13:65–82

Chapter 4
Fuzzy Logic

4.1 Logic and Fuzzy Logic: A Brief Overview

Fuzzy logic is a kind of reasoning inspired by the human reasoning. In a certain sense, it mimics the way humans perform decision making. It allows to handle uncertain propositions to obtain conclusions (Fig. 4.1).

Fuzzy logic represents the basis for approximate reasoning. Approximate reasoning (Zadeh 1975) refers to the process where conclusions are made from uncertain premises or hypotheses. The term fuzzy reasoning is also used, since the uncertainty is meant as fuzziness. This structure is similar to the one transferred by Socrates, but with imprecise information.

Example 4.1 Consider the following.

- Hairstylists stand up long;
- Standing up long frequently causes back pains;

Conclusion: Hairstyle jobs frequently cause back pains.

Long and frequently are imprecise terms and classical logic cannot handle sentences with such terms.

The approximate reasoning uses fuzzy sets, linguistic variables, different connectives, with several types of implication.

Classical logic connectives are

- **and** $\wedge$
- **or** $\vee$
- **not** $\neg$
- **implication** $\Longrightarrow$

S. Tomasiello et al., *Contemporary Fuzzy Logic*, Big and Integrated
Artificial Intelligence 1, https://doi.org/10.1007/978-3-030-98974-3_4

Fig. 4.1 Classical or fuzzy
logic?

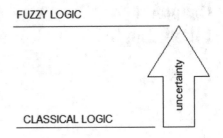

For instance:
If x is in **X and** y is in **Y or** z is in Z
then w is **not** in W.

The logical values for each connective are studied using truth tables (see Tables 4.1 and 4.2).

True sentences are assigned a logical value 1, while false sentences have a logical value 0 in classical logic (in the extension to the fuzzy case, truth values range between 0 and 1). The classical logic is sometimes called "two-valued logic". The classical logic connectives are briefly recalled:

- *connective* **and**

$$\wedge : \{0, 1\} \times \{0, 1\} \to \{0, 1\},$$

$$(u, v) \longmapsto \wedge(u, v) = u \wedge v = \min\{u, v\},$$

- *connective* **or**

$$\vee : \{0, 1\} \times \{0, 1\} \to \{0, 1\},$$

Table 4.1 Truth table of **and, or, implication**

u	v	$u \wedge v$	$u \vee v$	$u \Longrightarrow v$
1	1	1	1	1
1	0	0	1	0
0	1	0	1	1
0	0	0	0	1

Table 4.2 Truth table of **not**

u	$\neg u$
1	0
0	1

$$(u, v) \longmapsto \vee(u, v) = u \vee v = \max\{u, v\},$$

- *negation*

$$\neg : \{0, 1\} \rightarrow \{0, 1\}$$

$$u \longmapsto \neg v$$

(notice that $\neg u = 1 - u$),
- *implication*

$$\Longrightarrow : \{0, 1\} \times \{0, 1\} \rightarrow \{0, 1\}$$

$$(u, v) \longmapsto \Longrightarrow (u, v) = (u \Longrightarrow v).$$

4.2 Fuzzy Logic Basic Connectives

Let A be a fuzzy set. The aim is to determine how much the proposition "a belongs to A" is true. For the logical evaluation of the connectives in the fuzzy case, one has to extend the classical ones. These extensions are obtained by triangular norms and conorms (t-norms and t-conorms respectively). Given two fuzzy sets A and B, the t-norm is represented by the function

$$\underline{\wedge} : [0, 1] \times [0, 1] \rightarrow [0, 1], \tag{4.1}$$

which aggregates two membership grades.

A t-norm operator satisfies the following conditions for any $x, y, z \in [0, 1]$:

- $\underline{\wedge}(1, x) = 1\underline{\wedge}x = x$ (neutral element);
- $\underline{\wedge}(x, y) = x\underline{\wedge}y = y\underline{\wedge}x = \underline{\wedge}(y, x)$ (commutativity);
- $x\underline{\wedge}(y\underline{\wedge}z) = (x\underline{\wedge}y)\underline{\wedge}z$ (associativity);
- if $x \leq v$ and $y \leq w$, then $x\underline{\wedge}y \leq v\underline{\wedge}w$ (monotonicity).

This operator extends the operator $\wedge$ modelling the connective **and**.

The t-norm has to be regarded as a general class of intersection operators for fuzzy sets. Thanks to associativity, it is possible to compute the membership values for the intersection of more than two fuzzy sets by recursively applying a t-norm operator. There are different types of t-norms, namely (Fig. 4.2):

- $\underline{\wedge}_1(x, y) = \min\{x, y\} = x \wedge y$ (minimum-operator);
- $\underline{\wedge}_2(x, y) = xy$ (algebraic product);
- $\underline{\wedge}_3(x, y) = \max\{0, x + y - 1\}$ (bounded product);
- $\underline{\wedge}_4(x, y) = \begin{cases} x, & if \quad y = 1 \\ y, & if \quad x = 1 \\ 0, & otherwise. \end{cases}$ (drastic product).

Fig. 4.2 Fuzzy sets (top) and t-norm (bottom—minimum, continuous line; product, dashed line)

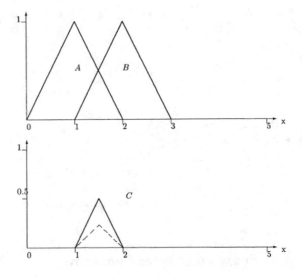

The Scilab function `tn=t_norm(A,B,s)` in the Appendix implements the formulas above. The minimum t-norm, algebraic product, bounded product and drastic product can be selected, by setting s, in the order, equal to 1, 2, 3 or any other value, This is obtained by means of the `case` instruction:

```
case 1 then
tn=min(A,B);
case 2
tn=A.*B;
case 3 tn=max(0,(A+B-1));
else tn = zeros(A);
tn(find(B==1)) = A(find(B==1));
tn(find(A==1)) = B(find(A==1));
end
```

A t-conorm (or s-norm) operator $\overline{\vee} : [0, 1] \times [0, 1] \rightarrow [0, 1]$ satisfies the following conditions for any $x, y, z \in [0, 1]$:

- $\overline{\vee}(0, x) = 0 \overline{\vee} x = x$ (neutral element);
- $\overline{\vee}(x, y) = x \overline{\vee} y = y \overline{\vee} x = \overline{\vee}(y, x)$ (commutativity);
- $x \overline{\vee}(y \overline{\vee} z) = (x \overline{\vee} y) \overline{\vee} z$ (associativity);
- if $x \leq v$ and $y \leq w$, then $x \overline{\vee} y \leq v \overline{\vee} w$ (monotonicity).

This operator extends the operator $\vee$ modelling the connective **or**. The class of t-conorm operators is referred to the union of fuzzy sets.

Types of t-conorms:

- $\overline{\vee}_1 = \max\{x, y\} = x \vee y$ (max-operator);
- $\overline{\vee}_2 = x + y - xy$ (algebraic sum);
- $\overline{\vee}_3 = \min\{1, x + y\}$ (bounded sum);

Fig. 4.3 Some t-norms: minimum (red), algebraic product (light blue), Schweizer and Sklar's, with p=1 (purple) and p=10 (yellow)

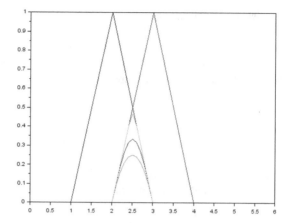

$$\bullet \ \overline{\vee}_4 = \begin{cases} x, & if \ \ y = 0 \\ y, & if \ \ x = 0 \ \ (\text{drastic sum}) \\ 1, & otherwise. \end{cases}$$

The Scilab function `tn=t_conorm(A,B,s)` is very similar to the function implementing the t-norm formulas. The code listing is in the Appendix.

There exist parameterized t-norms and t-conorms. They represent a generalization of t-norm and t-conorm by means of a parameter p.

For instance, Schweizer and Sklar's:

$$\triangle_S(x, y, p) = [\max\{0, (x^{-p} + y^{-p} - 1)\}]^{-1/p}, \tag{4.2}$$

$$\overline{\vee}_S(x, y, p) = 1 - [\max\{0, ((1 - x)^{-p} + (1 - y)^{-p}) - 1]^{-1/p}, \tag{4.3}$$

with $p \in (0, \infty)$. It has been proved that

$$\lim_{p \to 0} \triangle_S(x, y, p) = xy, \tag{4.4}$$

$$\lim_{p \to \infty} \triangle_S(x, y, p) = \min(x, y). \tag{4.5}$$

and this can be seen in Fig. 4.3.

An operator $\Rightarrow: [0, 1] \times [0, 1] \to [0, 1]$ is a fuzzy implication if it

- reproduces the classical implication table;
- is decreasing in the first variable;
- is increasing in the second variable.

A map $v : [0, 1] \to [0, 1]$ is a negation if it satisfies the following conditions

- $v(0) = 1, v(1) = 0$ (boundary conditions);
- v is decreasing (monotonicity).

Besides, if v is strictly decreasing and the boundary conditions hold, as well as the involution condition $v(v(x)) = x$, then v is called a strong negation. For instance, the map $v_1(x) = 1 - x$ is a strong negation.

The t-norm and the t-conorm are said to be dual with respect to a negation v, if they satisfy the De Morgan's laws

$$v(x \underline{\wedge} y) = v(x) \overline{\vee} v(y), \tag{4.6}$$

$$v(x \overline{\vee} y) = v(x) \underline{\wedge} v(y), \tag{4.7}$$

for all pairs $(x, y) \in [0, 1] \times [0, 1]$.

The fuzzy implications represent a class of fuzzy relations.

There are many fuzzy implication membership functions in literature. In general, two interpretations are possible.

- $A \rightarrow B$ interpreted as A **coupled with** B ("association" of A and B)

$$R = A \rightarrow B = A \times B, \tag{4.8}$$

where $A \times B$ is the Cartesian product of the fuzzy sets A and B, with the membership function

$$\mu_R(x, y) = \underline{\wedge}[\mu_A(x), \mu_B(y)], \qquad \forall x \in X, y \in Y, \tag{4.9}$$

where $\underline{\wedge}$ is a t-norm. In this group, there are:

– Mamdani

$$\mu_R(x, y) = \min\{\mu_A, \mu_B\}; \tag{4.10}$$

– Larsen

$$\mu_R(x, y) = \mu_A \mu_B; \tag{4.11}$$

– Bounded product

$$\mu_R(x, y) = \max\{0, \mu_A(x) + \mu_B(y) - 1\}. \tag{4.12}$$

- $A \rightarrow B$ interpreted as A **entails** B and a possible formula is

$$\mu_R(x, y) = \overline{\vee}[\mu_{\overline{A}}(x), \mu_B(y)], \qquad \forall x \in X, y \in Y, \tag{4.13}$$

where $\overline{\vee}$ is a t-conorm and $\overline{A} = 1 - A$ is the complement of A. Some examples:

– Kleene-Dienes

$$\mu_R(x, y) = \max\{1 - \mu_A(x), \mu_B(y)\}; \tag{4.14}$$

– Lukasiewicz

$$\mu_R(x, y) = \max\{1, 1 - \mu_A(x) + \mu_B(y)\}. \tag{4.15}$$

4.3 Generalized Composition of Fuzzy Relations

The min operator of the max-min composition can be replaced by any t-norm.

Definition 4.1 The *sup-t* composition of two fuzzy relations R over $X \times Y$ and S over $Y \times Z$ is a fuzzy relation whose membership function is given by

$$\mu_{R \circ^t S}(x, z) = \sup_{y \in Y} \triangle[\mu_R(x, y), \mu_S(y, z)], \qquad \forall (x, z) \in X \times Z. \tag{4.16}$$

For finite sets, it is *sup = max*. For the product-type t-norm, there is the *max-product* composition.

Definition 4.2 The *inf-c* composition of two fuzzy relations R over $X \times Y$ and S over $Y \times Z$ is a fuzzy relation whose membership function is given by

$$\mu_{R \circ_c S}(x, z) = \inf_{y \in Y} \overline{\triangledown}[\mu_R(x, y), \mu_S(y, z)], \qquad \forall (x, z) \in X \times Z. \tag{4.17}$$

For the maximum t-conorm, one has the *inf-max* composition.

4.4 Linguistic Variables

Definition 4.3 A linguistic variable x in the universe X is a variable whose values are fuzzy sets of X.

A linguistic variable is characterized by a quintuple $(x, T(x), X, S, M)$ in which

- x is the name of the variable;
- $T(x)$ is the term set of x;
- X is the universe of the discourse;
- S is a syntactic rule which generates the terms in $T(x)$;
- M is a semantic rule which associates with each linguistic value a its meaning $M(a)$, where $M(a)$ denotes a fuzzy set in X.

Notice that the term set consists of several *primary terms* (e.g. young, old) modified by *negation* (not) and/or *linguistic modifiers*, also called *hedges* (very, more or less, quite).

Definition 4.4 A term set $T = \{t_1, t_2, \ldots, t_n\}$ of a linguistic variable is orthogonal if the following holds

$$\sum_{i=1}^{n} \mu_{t_i}(x) = 1, \qquad \forall x \in X \tag{4.18}$$

where t_i's are convex and normal fuzzy sets.

Example 4.2 Let x be a linguistic variable, labelled as "age", with $X = [0, 100]$ defining the universe and x being the age in years. A term set of the variable x could be (Fig. 4.4):

T(age)={very young, young, middle-aged, not young and not old, old, very old}.
Every term is a fuzzy set. For instance:

$$M(old) = \{(x, \mu_{old}(x)) | x \in [0, 100]\},$$

where $\mu_{old}(x)$ is the membership function. The syntactic rule refers to the way the linguistic values in the term set T(age) are generated. The semantic rule defines the membership function of each linguistic value of the term set.

"Age is young" represents the assignment of the linguistic value "young" to the linguistic variable *age*. When age is interpreted as a numerical variable, then "age=21", for instance, represents the assignment of the numerical value 21 to the variable.

4.5 Linguistic Modifiers and Composite Linguistic Terms

Linguistic modifiers are used to change attributes, that is, modelling adverbs.

Let $\mathcal{F}(X)$ denote the family of all fuzzy sets defined in the universe of discourse X.

Definition 4.5 A fuzzy modifier m over X is a map defined on $\mathcal{F}(X)$ with values on $\mathcal{F}(X)$:

$$m : \mathcal{F}(X) \to \mathcal{F}(X).$$

The main fuzzy modifiers are

- expansive if, for all $A \in \mathcal{F}(X)$, $A \subseteq m(A)$, that is, $\mu_A(x) \leq \mu_{m(A)}(x)$;
- restrictive if, for all $A \in \mathcal{F}(X)$, $A \supseteq m(A)$, that is, $\mu_A(x) \geq \mu_{m(A)}(x)$.

The most commonly used fuzzy modifiers are power type. A modifier is power type if for each $A \in \mathcal{F}(X)$ we have

$$m_s(A) = (A)^s,$$

that is,

$$\mu_{m(A)}(x) = (\mu_A(x))^s,$$

for some $s \in [0, \infty)$. Notice that if $s < 1$ then m_s is expansive and if $s > 1$ then m_s is restrictive, since $\mu_A(x) \in [0, 1]$.

Let A be a linguistic variable characterized by a fuzzy set with a membership function $\mu_A(x)$, $x \in X$. The following operation is called *concentration*

$$CON(A) = A^2,$$

while the *dilation* is defined as

$$DIL(A) = A^{0.5}.$$

It is assumed that $CON(A)$ and $DIL(A)$ express the linguistic modifiers "very" and "more or less" respectively. It is possible to combine linguistic terms by using hedges and operators such as not, and. For instance, by assuming that the linguistic terms "young" and "old" are defined by means of some membership functions, we have

- more or less young: $DIL(young) = young^{0.5}$,
- not young and not old: $\neg young \wedge \neg old$,
- old but not too old: $old \wedge \neg old^2$,
- extremely young: $CON(CON(CON(young))) = ((young^2)^2)^2$,

considering that extremely can be regarded as very very very.

Fig. 4.4 Primary and composite linguistic values of the variable "age"

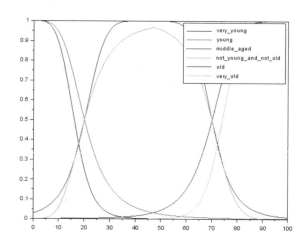

4.6 Fuzzy Propositions, Modus Ponens and Generalized Modus Ponens

A fuzzy IF-THEN rule, also known as fuzzy implication, fuzzy rule, or fuzzy proposition, assumes the form

$$\text{IF } (x \text{ is } A) \text{ THEN } (y \text{ is } B),$$

where A and B are linguistic terms, represented by fuzzy sets $A \in \mathcal{F}(X)$ and $B \in \mathcal{F}(Y)$, describing the variables x and y, respectively. The IF-THEN rule is often denoted as $A \rightarrow B$.

In the fuzzy rule above, "x is A" is called the antecedent or premise, while "y is B" is called the consequent or conclusion. The numerical value associated with "x is A" represents the state, while the numerical value "y is B" represents the answer. The fuzzy IF-THEN rule above can be represented by a binary fuzzy relation $R = A \rightarrow B$ in the product space $X \times Y$

$$R = A \rightarrow B = \{((x, y), \mu_{R=A \rightarrow B}(x, y) | (x, y) \in X \times Y\}. \tag{4.19}$$

The fuzzy propositions appear in the fuzzy modus ponens.
Let us recall the classical modus ponens:

$$(A \wedge (A \implies B)) \implies B, \tag{4.20}$$

which can be interpreted as: "If A is true and if the statement 'If A is true then B is true' is also true, then B is true."

In the fuzzy modus ponens, fuzzy propositions appear. They are formulated by means of linguistic variables and their attributes. For instance: *error* (linguistic variable) is *small* (attribute).

Let A be a fuzzy set modelling a linguistic term. The logical value of "$\bar{x}$ *is* A" is the number $\mu_A(\bar{x})$, that indicates how much $\bar{x}$ is in agreement with A.

Like the classical one, the fuzzy modus ponens consists of

- *fact* or *premise* (x is A),
- *implication* (If x is A then y is B),
- *conclusion* (y is B).

For instance, consider the following.
Premise: This tomato is red
Implication: If the tomato is red then the tomato is ripe
Conclusion: This tomato is ripe

The classical modus ponens can be given by the max-min composition rule $B = A \circ R$, where the relation R is obtained by a fuzzy implication that models the conditional sentence "If x is A then y is B".

Let A, A', B, B' be fuzzy sets, then the generalized modus ponens reads

Premise: x is A'

Implication: if x is A, then y is B

Conclusion: y is B'

Example 4.3 Consider the following generalized modus ponens.

Premise: This tomato is more or less red

Implication: If a tomato is red then it is ripe

Conclusion: This tomato is more or less ripe

The dominant wavelength of red is approximately 620–720 nm. A continuous fuzzy set to model redness of the tomato could be a trapezoidal one. Let $X = [620, 720]$ be the universe and $A = (620, 645, 690, 720)$ be such a fuzzy set.

Applying the modifier: `more_or_less_red=red^0.5`.

To model ripeness of tomato, one can consider vitamin C content of tomatoes (Valsikova et al. 2017). Ripe tomatoes have around 16–19 mg of vitamin C. Let $Y = [16, 19]$ be the universe and $B = (16, 17, 19, 19)$ be the fuzzy set modelling ripe, which is assumed to be a right trapezoid.

Let us suppose the following fuzzy relation matrix to model the implication:

$$R = \begin{bmatrix} 1 & 0.25 & 0.1 & 0.7 \\ 0.85 & 1 & 0.15 & 0 \\ 0.1 & 0.88 & 1 & 0.35 \\ 0.7 & 0 & 0.95 & 1 \end{bmatrix}.$$

The universes of the two continuous fuzzy sets red and ripe, X and Y respectively, could be normalized, so that $\bar{x} = (x - 620)/(720 - 620)$ and $\bar{y} = (y - 16)/(19 - 16)$. Then, the new universes would be $\bar{X} = \bar{Y} = [0, 1]$.

By using Scilab, the modelled fuzzy sets and the result are shown in Fig. 4.5.

Problems

1. Consider the linguistic variable "skin". Elaborate on the term set and plot the related membership functions.
2. Apply a few linguistic modifiers to some of the attributes of the variable "skin" and plot the new membership functions.
3. Consider the linguistic variable "light intensity". Choose an orthogonal term set. Apply some linguistic modifiers and check whether the term set is still orthogonal.
4. Show an example of generalized modus ponens in Scilab.

Fig. 4.5 Generalized modus
ponens in the Scilab console

```
--> x1=[0.025 0.077 0.31 0.85];

-->

--> red=tra(0,0.25,0.7,1,x1)
red =

  0.1  0.308  1.  0.5


--> more_or_less_red=red^0.5
more_or_less_red  =

  0.3162278  0.5549775  1.  0.7071068


--> R=[1 0.25 0.1 0.7;
>   0.85 1 0.15 0;
>   0.1 0.88 1 0.35;
>   0.7 0 0.95 1];

-->

-->   more_or_less_ripe=max_min(more_or_less_red,R)
more_or_less_ripe  =

  0.7  0.88  1.  0.7071068
```

References

Valsikova M, Komár P, Rehus M (2017) The effect of varieties and degree of ripeness to vitamin6
   C content in tomato fruits. Acta Horticulturae et Regiotecturae 20. 10.1515/ahr-2017-0010
Zadeh LA (1975) The concept of a linguistic variable and its application to approximate reasoning,
   I, II, III. Inform Sci 8:199–249, 301–357, 43–80
Zimmermann H-J (2001) Fuzzy set theory and its applications. Kluwer Academic Publishers

Chapter 5
Fuzzy Inference Systems

5.1 Fuzzy Rule Bases

The terms "fuzzy inference system", "fuzzy rule-based system", "fuzzy controller" are used in the literature without distinction to mean a computing framework based on concepts of fuzzy set theory, fuzzy if-then rules and fuzzy reasoning.

A fuzzy inference system (FIS) can take either fuzzy or non-fuzzy inputs. In general, the outputs may be fuzzy sets. Then a method of defuzzification is needed in order to extract a numerical value that best represents a fuzzy set. Figure 5.1 shows a general FIS scheme. Its modules are described in the following.

- **Fuzzification module.** The inputs of the system are modelled by means of fuzzy sets. The membership functions are usually chosen by consulting the experts in the domain. For each input numerical value, one obtains a vector whose elements are the membership degrees in the considered fuzzy set.
- **Rule-Base module.** It consists of some fuzzy propositions, such as

$$\text{If } x_1 \text{ is } A_1 \text{ and } x_2 \text{ is } A_2 \text{ and } \dots \text{ and } x_n \text{ is } A_n$$

$$\text{Then } u_1 \text{ is } B_1 \text{ and } u_2 \text{ is } B_2 \text{ and } \dots \text{ and } u_m \text{ is } B_m$$

The general form of a fuzzy rule base (FRB) is
R_1: *Fuzzy proposition 1*
or
R_2: *Fuzzy proposition 2*
or ...
R_r: *Fuzzy proposition r*
Thanks to the information gathered from the experts of the domain, the input and output variables with the term sets are defined, by choosing the suitable membership functions. Methods to obtain the membership functions are curve fitting, interpolation and neural networks-like.
- **Fuzzy inference module.** It translates fuzzy propositions into fuzzy relations. In this module, the kind of t-norms, t-conorms and fuzzy implications, to obtain the fuzzy relation that models the rule base, are fixed.

© The Author(s), under exclusive license to Springer Nature Switzerland AG 2022
S. Tomasiello et al., *Contemporary Fuzzy Logic*, Big and Integrated
Artificial Intelligence 1, https://doi.org/10.1007/978-3-030-98974-3_5

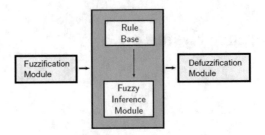

Fig. 5.1 The general FIS scheme

- **Defuzzification module.** It allows to represent a fuzzy set by a numerical value (real number). There are several defuzzification methods (see Sect. 5.3.1).

5.1.1 Single Rule with Single Antecedent

Let us consider the generalized modus ponens (GMP)

Premise: x is A*
Rule: IF x is A THEN y is B
Conclusion: y is B*
By using the max-min composition and the Mamdani's implication:

$$\mu_{B^*}(y) = \max_x \min[\mu_{A^*}(x), \mu_R(x, y)] = \max_x \min[\mu_{A^*}(x), \min[\mu_A(x), \mu_B(y)]], \tag{5.1}$$

that is

$$\mu_{B^*}(y) = \min[\omega, \mu_B(y)], \tag{5.2}$$

where $\omega = \max_x \min[\mu_{A*}(x), \mu_A(x)]$.

Intuitively, ω can be meant as a measure of the degree of belief for the antecent. The resulting degree of belief for the consequent (B*) should be no greater than ω, as depicted in Fig. 5.2.

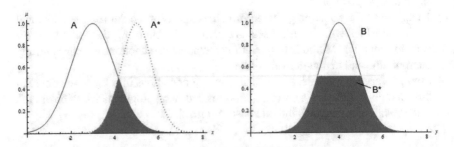

Fig. 5.2 Graphic interpretation of GMP by using Mamdani's implication

5.1.2 Single Rule with Multiple Antecedents

Let us consider the following generalized modus ponens

Premise: x is A* and y is B*
Rule: IF x is A and y is B THEN z is C
Conclusion: z is C*
The fuzzy rule can be put in the form $A \times B \rightarrow C$ and the resulting C* is

$$C^* = (A^* \times B^*) \circ (A \times B \rightarrow C).$$

Hence

$$\mu_{C^*}(z) = \max_{x,y} \min[\min[\mu_{A^*}(x), \mu_{B^*}(y)], \min[\mu_A(x), \mu_B(y), \mu_C(z)]] =$$

$$= \min[\max_{x,y} \min[\mu_{A^*}(x), \mu_{B^*}(y), \mu_A(x), \mu_B(y)], \mu_C(z)] =$$

$$= \min[\max_x \min[\mu_{A^*}(x), \mu_A(x)], \max_y \min[\mu_{B^*}(y), \mu_B(y)], \mu_C(z)].$$

Let $\omega_1 = \max_x \min[\mu_{A^*}(x), \mu_A(x)]$ and $\omega_2 = \max_y \min[\mu_{B^*}(y), \mu_B(y)]$ denote the *degrees of compatibility* between A and A*, and B and B*, respectively. Then

$$\mu_{C^*}(z) = \min[\sigma, \mu_C(z)]. \tag{5.3}$$

$\sigma = \min(\omega_1, \omega_2)$ is called *firing strength* or *degree of fulfillment* of the fuzzy rule, representing the degree to which the antecedent is satisfied.

It is possible to prove the following decomposition rule (e.g. see Jang et al. 1997):

$$C^* = C_1^* \cap C_2^*, \tag{5.4}$$

where $C_1^* = A^* \circ (A \rightarrow C)$ and $C_2^* = B^* \circ (B \rightarrow C)$, each one representing the inferred fuzzy set of a GMP problem for a single fuzzy rule with a single antecedent.

5.1.3 Multiple Rules with Multiple Antecedents

Premise: x is A^* and y is B^*

Rule 1: IF x is A_1 and y is B_1 THEN z is C_1
Rule 2: IF x is A_2 and y is B_2 THEN z is C_2
Conclusion: z is C^*
Let $R_1 = A_1 \times B_1 \rightarrow C_1$ and $R_2 = A_2 \times B_2 \rightarrow C_2$. Recalling that the max-min composition operator $\circ$ is distributive over the $\cup$ operator then

$$C^* = (A^* \times B^*) \circ (R_1 \cup R_2) = C_1^* \cup C_2^*,$$

where $C_1^* = (A^* \times B^*) \circ R_1$ and $C_2^* = (A^* \times B^*) \circ R_2$ are the inferred fuzzy sets for the implications (or rules) 1 and 2 respectively.

At this point, it is possible to summarize the approximate reasoning by four steps.

- *Degrees of compatibility*, to be determined by comparing the facts with the antecedents of the fuzzy rules.
- *Firing strength*, obtained combining degrees of compatibility; it indicates the degree to which the antecedent of the rule is satisfied.
- *Qualified consequent*, generated by applying the firing strength to the consequent of a rule.
- *Overall output*, by aggregating all the qualified consequents.

5.2   Fuzzy Controller

A fuzzy controller is a typical case of FRB system. It consists of a fuzzy system, where the inputs represent "conditions" while the outputs are "actions". Hence, each rule can be meant as

IF "condition" THEN "action".

Perhaps, control is the main application of fuzzy logic. The first relevant article in the control field using fuzzy logic was by Mamdani Assilian (1975). They investigated the control of a steam engine. Their idea lay on the fact that human operators express control strategies in linguistic form and not in a mathematical way. Since then, fuzzy controllers have become very popular and used in different types of electrical appliances. As a general design criterion, the input fuzzy sets must cover the universes of discourse so that any value of input variables will produce at least one non-zero membership value. The number of input fuzzy sets, their linguistic names and shapes are design parameters determined by the developer on the basis of the characteristics of the system to be controlled, own knowledge and experience. Usually, 2–13 fuzzy sets are used for each input variable and a larger number is uncommon. Different numbers of fuzzy sets may be used for different input variables. Each fuzzy set is assigned a different linguistic name. Some common names are "Negative Large" (NL)," "Negative Medium" (NM), "Negative Small" (NS), "Approximately Zero" (Z), "Positive Small" (PS), "Positive Medium" (PM) and "Positive Large" (PL). Regarding the rule base, there are no systematic tools for forming it. Usually, intuitive knowledge and experience (of an operator) are employed, so that the fuzzy controller is designed as a simple expert system.

Fuzzy control generates a nonlinear mapping from the input variable space to the output variable space. Thus, fuzzy control can be regarded as nonlinear control. Its big advantage is that it provides an effective and efficient methodology without using highly advanced mathematics. Unlike the traditional controller design methodology, explicit system model is not needed by fuzzy control. System model is implicitly built into fuzzy rules, fuzzy logic operation and fuzzy sets in a vague manner. A disadvantage is that a fuzzy controller usually has more design parameters than a comparable conventional controller.

5.2.1 Fuzzy Controllers and Dynamical Systems

In the theory of non-fuzzy discrete-time systems, a system is defined through its state equations:

$$x(t + 1) = f(x(t), u(t)), \tag{5.5}$$
$$y(t) = g(x(t), u(t)), \tag{5.6}$$

where $u(t), y(t), x(t)$ denotes the input, the output and the state at time t, respectively, while f and g are mappings from $X \times U$ to X and Y, respectively. When applying fuzzy controllers to dynamical systems, the inputs are the state variables whereas the outputs are the state variations. From the modelling perspective, such a procedure is reasonable when the available information is incomplete and the phenomenon is partially understood.

Let $y(t)$ be the system output, for any positive integer t. Let $S(t)$ denote the desired system output trajectory. It is usual to employ the error $e(t)$ and change of error (or rate) $r(t)$ as input variables:

$$e(t) = S(t) - y(t), \qquad r(t) = e(t) - e(t - 1) = y(t - 1) - y(t). \tag{5.7}$$

Assuming that scaling factors for error and rate are K_e and K_r, respectively, then the scaled error and rate are

$$E(t) = K_e e(t), \qquad R(t) = K_r r(t). \tag{5.8}$$

Two arrays of fuzzy sets are needed: one for $E(t)$ and the other for $R(t)$. The use of "Positive" and "Negative" in the linguistic names is necessary because $e(t)$ and $r(t)$ can be positive and negative. Figure 5.3 shows a typical fuzzy partition for the variable error, obtained by using the Scilab Fuzzy Logic Toolbox.

5.3 Mamdani System

The Mamdani inference method uses

- the minimum t-norm to model the fuzzy implication;
- the minimum t-norm for the logical connective "and";

Fig. 5.3 Typical fuzzy sets
for the input variable error

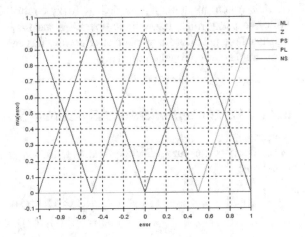

- the maximum t-conorm for the logical connective "or", which connects the fuzzy rules of the rule base.

In literature, there are some variants using different t-norm and t-conorm operators.

For instance, let us consider a Mamdani inference system with three inputs and three rules.

Premise: x_1 is A_1 and x_2 is A_2 and x_3 is A_3
Rule 1: If x_1 is A_{11} and x_2 is A_{12} and x_3 is A_{13} then y is B_1
Rule 2: If x_1 is A_{21} and x_2 is A_{22} and x_3 is A_{23} then y is B_2
Rule 3: If x_1 is A_{31} and x_2 is A_{32} and x_3 is A_{33} then y is B_3
Conclusion: y is B

Let us recall that $B = \overline{B}_1 \cup \overline{B}_2 \cup \overline{B}_3$, where $\overline{B}_1 = (A_1 \times A_2 \times A_3) \circ R_1$, $\overline{B}_2 = (A_1 \times A_2 \times A_3) \circ R_2$ and $\overline{B}_3 = (A_1 \times A_2 \times A_3) \circ R_3$, with $R_1 = A_{11} \times A_{12} \times A_{13} \rightarrow B_1$, $R_2 = A_{21} \times A_{22} \times A_{23} \rightarrow B_2$ and $R_3 = A_{31} \times A_{32} \times A_{33} \rightarrow B_3$.

Let $\omega_{ij} = \max_{x_i} \min[\mu_{A_i}(x_i), \mu_{A_{ij}}(x_i)]$ denote the *degrees of compatibility* between A_i and A_{ij}, with $i, j = 1, 2, 3$. Then

$$\mu_{\overline{B}_i}(y) = \min(\sigma_i, \mu_{B_i}(y)), \tag{5.9}$$

and

$$\mu_B = \max(\mu_{\overline{B}_1}(y), \mu_{\overline{B}_2}(y), \mu_{\overline{B}_3}(y)). \tag{5.10}$$

5.3.1 Defuzzification Methods

The operation of defuzzification can be regarded as a function mapping fuzzy sets of the real numbers into real numbers. There are different defuzzification methods (Fig. 5.4).

- **Centroid or center of gravity.** It is a weighted average, i.e.

$$G(B) = \frac{\int_Y y\mu_B(y)dy}{\int_Y \mu_B(y)dy}, \qquad (5.11)$$

and in the discrete case

$$G(B) = \frac{\sum_{i=1}^n y_i \mu_B(y_i)}{\sum_{i=1}^n \mu_B(y_i)}. \qquad (5.12)$$

- **Center of area (bisector of area).** In the center of area (CoA), the defuzzified value is the support element that divides the area under the curve of a continuous membership function into two equal parts

$$\int_{y_{min}}^{d_{CoA}} \mu_B(y)dy = \int_{d_{CoA}}^{y_{max}} \mu_B(y)dy. \qquad (5.13)$$

- **Center of maximum.** It takes into account just the regions of major possibility among the allowed values of the considered variable. Let $C_m = \{y|\mu_B(y) = hgt(B)\}$, where $hgt(B)$ is the height of B. Then the center of maximum is

$$C(B) = \frac{\min C_m + \max C_m}{2}. \qquad (5.14)$$

- **Mean of maximum.** It is particularly used for discrete fuzzy sets,

$$M(B) = \sum_{i=1}^K \{y_i|y_i \in C_m\}/K, \qquad (5.15)$$

where K is the total number of elements in the set C_m.

Example 5.1 Let us consider the trapezoidal fuzzy number $A = (0, 1, 2, 4)$ and compute its centroid. The left and right side of A are $y = f_1(x) = x$ and $y = f_2(x) = (4 - x)/2$. Then the numerator of $G(A)$ is

$$\int_0^1 xf_1(x)dx + \int_1^2 xdx + \int_2^4 xf_2(x)dx = 1/3 + 3/2 + 8/3 = 9/2. \qquad (5.16)$$

The area of the trapezoid is $5/2$. Hence, $G(A) = 9/5$.

Example 5.2 The center of maximum for the fuzzy number $A = (a_1, a_2, a_3)$ is $C(A) = a_2$. For the fuzzy number $A = (a_1, a_2, a_3, a_4)$, it is $C(A) = (a_2 + a_3)/2$, because $C_m = [a_2, a_3]$ is the core of A.

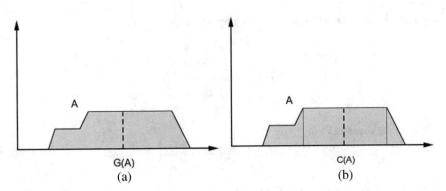

Fig. 5.4 **a** Centroid **b** Center of maximum

Example 5.3 For the finite fuzzy number

$$A = \{(0, 0), (2, 0.8), (4, 0.6), (6, 0.8), (4, 0.3), (5, 0.1)\},$$

the center of maximum $C(A) = (2 + 6)/2 = 4$. When the result does not equal a member of the support of A, then rounding up or down is required. For example, if the second pair of A had been $(1, 0.8)$, then $C(A) = (1 + 6)/2 = 3.5$, which could be rounded to 4.

5.4 Takagi-Sugeno-Kang System

The main difference between the Takagi–Sugeno–Kang (TSK) inference method and Mamdani's is in the type of consequent of each rule. In the TSK method, the consequent is given by a function of the input values of the rule. Hence, the generic rule R_k is written as

R_k: If x_1 is A_{k1} and x_2 is A_{k2}... and x_n is A_{kn} then y is $y_k = g_k(x_1, \ldots, x_n)$,

with $k = 1, \ldots, r$. The final output is given by

$$y = \frac{\sum_{j=1}^{r} \overline{\omega}_j g_j(x_1, x_2, \ldots, x_n)}{\sum_{j=1}^{r} \overline{\omega}_j}, \tag{5.17}$$

with the weights $\overline{\omega}_j = \mu_{A_{j1}}(x_1) \triangle \mu_{A_{j2}}(x_2) \triangle \ldots \triangle \mu_{A_{jn}}(x_n)$, for any t-norm $\triangle$.

When the functions g_j are linear, then the system is called Takagi–Sugeno (TS).

Example 5.4 Let us consider a TSK system with the following rules:

R_1: If x_1 is A_{11} and x_2 is A_{12} and x_3 is A_{13} then y is $y_1 = g_1(x_1, x_2, x_3)$,

or

R_2: If x_1 is A_{21} and x_2 is A_{22} and x_3 is A_{23} then y is $y_2 = g_2(x_1, x_2, x_3)$.

or

R_3: If x_3 is A_{31} and x_2 is A_{22} and x_3 is A_{33} then y is $y_3 = g_3(x_1, x_2, x_3)$.

Then the final output, by using the minimum t-norm, is

$$y = \frac{\overline{\omega}_1 y_1 + \overline{\omega}_2 y_2 + \overline{\omega}_3 y_3}{\overline{\omega}_1 + \overline{\omega}_2 + \overline{\omega}_3},$$

with $\overline{\omega}_i = \min[\mu_{A_{i1}}(x_1), \mu_{A_{i2}}(x_2), \mu_{A_{i3}}(x_3)]$, for $i = 1, 2, 3$, representing the degree of fulfillment of the ith fuzzy rule.

Example 5.5 Let us model a simple heating system by using a Mamdani FIS. The input variables are *temperature* and *change of temperature*. The output variable is *power*. The term set of temperature is:

$$T(\text{temperature}) = \{\text{low, comfortable, high}\},$$

while the term set of change of temperature is:

$$T(\text{change of temperature}) = \{\text{negative big, negative small, zero, positive small, positive big}\},$$

For power it is:

$$T(\text{power}) = \{\text{small, medium, big}\}.$$

The FIS has been implemented by using the Scilab Fuzzy Logic Toolbox (see how to install and use it in the Appendix). Figures 5.5 and 5.6 show the input and output variables, respectively.

The rule base is:

R1: IF temperature IS low AND change of temperature IS negative small THEN power IS medium

R2: IF temperature IS low AND change of temperature IS zero THEN power IS medium

R3: IF temperature IS low AND change of temperature IS positive small THEN power IS small

R4: IF temperature IS comfortable AND change of temperature IS negative small THEN power IS medium

R5: IF temperature IS comfortable AND change of temperature IS zero THEN power IS small

R6: IF temperature IS comfortable AND change of temperature IS positive small THEN power IS small

R7: IF temperature IS high AND change of temperature IS negative big THEN power IS medium

R8: IF temperature IS low AND change of temperature IS negative big THEN power IS big

R9: IF temperature IS low AND change of temperature IS positive big THEN power IS small

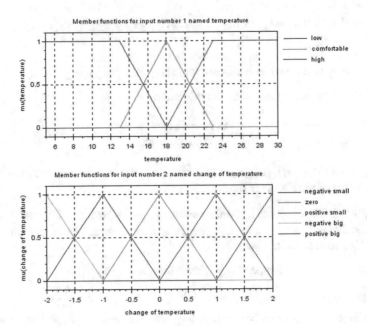

Fig. 5.5 Linguistic input variables for the heating system

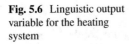

Fig. 5.6 Linguistic output
variable for the heating
system

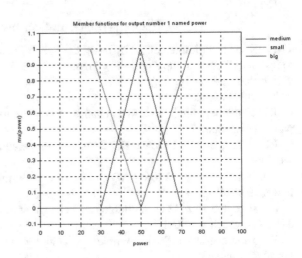

Let us assume that the current temperature is 21 °C and that the change of temperature is −0.5 °C/min.

This temperature is "comfortable" with degree 0.4 and "high" with degree 0.6.

The change of temperature is "negative small" with degree 0.5 and "zero" with degree 0.5.

Two rules are activated:

R4- IF temperature IS comfortable AND change of temperature IS negative small THEN power IS medium

R5- IF temperature IS comfortable AND change of temperature IS zero THEN power IS small

Then for each activated rule, there is the same degree of firing

$$\sigma = \min\{0.4, 0.5\} = 0.4.$$

The consequents of the rules are determined by projection onto the relative membership functions; symbolically

$$\mu_4(y) = \min\{\sigma, \mu_M\}, \mu_5(y) = \min\{\sigma, \mu_S\},$$

where μ_M and μ_S are the membership functions of the fuzzy sets medium and small. The resulting output, without defuzzification, (Fig. 5.7) is obtained by computing the maximum

$$\overline{\mu} = \max\{\mu_4(y), \mu_5(y)\}.$$

The defuzzified output by the centroid method is 32.789. By the mean of maximum it turns out to be 31, while it is 33 by the bisector method.

Example 5.6 Let us consider a TSK FIS whose inputs are the values of an unknown function f at time $t - 1$ and t, i.e. $f(t - 1)$ and $f(t)$ respectively, and whose output is the value of the same function at time $t + 1$, i.e. $f(t + 1)$. For each input variable there are two terms: "small" and "large". There are four rules

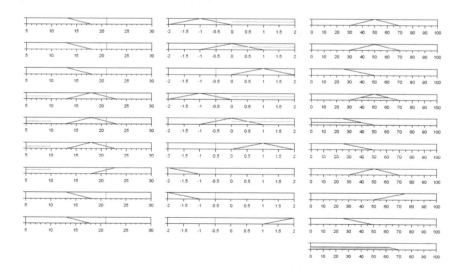

Fig. 5.7 Rules activation and final fuzzy output (in grey) for temperature $21\,^\circ$C and change of temperature $-0.5\,^\circ$C/min

R1: IF f(t-1) IS small AND f(t) IS small THEN f(t+1) IS g1
R2: IF f(t-1) IS small AND f(t) IS large THEN f(t+1) IS g2
R3: IF f(t-1) IS large AND f(t) IS large THEN f(t+1) IS g3
R4: IF f(t-1) IS large AND f(t) IS small THEN f(t+1) IS g4

where

$$g_1 = 0.5 f(t-1) + f(t) + 0.6,$$

$$g_2 = 0.3 f(t-1) + 0.8 f(t) + 1,$$

$$g_3 = 0.9 f(t-1) + f(t) + 0.4,$$

$$g_4 = 0.8 f(t-1) + 0.7 f(t) + 0.9.$$

Figure 5.8 shows the chosen membership functions and the resulting FIS output as a function of the input values. Figure 5.9 shows a different choice for the membership functions, i.e. S-membership function, and the corresponding output. As one can see the latter turns out to be smoother.

Example 5.7 In this application example, the Mamdani system is used for breast cancer classification. To predict whether a tumour is malignant or not, 9 input features are considered, as detailed in Table 5.1

(Data source: http://archive.ics.uci.edu/ml/datasets/breast+cancer+wisconsin+ (diagnostic)).

From the original data set, consisting of 683 samples, 140 instances were randomly selected for testing. The data set was normalized in the range [0, 1], which represent the universe of discourse (UoD). For the sake of brevity, abbreviations were used for the input variables.

For each input variable, 3 terms were assigned, namely "low", "medium", "high". The output variable, here named "tumour", may assume the linguistic value "malignant" or "benign" (Fig. 5.10).

The rule base is

R1: IF CT IS low AND UCZ IS low AND UCS IS low AND MA IS low AND SECS IS low AND BN IS low AND BC IS low AND NN IS low AND MS IS low THEN tumour IS benign

R2: IF CT IS medium AND UCZ IS low AND UCS IS low AND MA IS low AND SECS IS low AND BN IS low AND BC IS medium AND NN IS medium AND MS IS low THEN tumour IS malignant

R3: IF CT IS high AND UCZ IS high AND UCS IS medium AND MA IS high AND SECS IS high AND BN IS high AND BC IS medium AND NN IS medium AND MS IS medium THEN tumour IS malignant

R4: IF CT IS high AND UCZ IS medium AND UCS IS medium AND MA IS medium AND SECS IS medium AND BN IS low AND BC IS medium AND NN IS low AND MS IS low THEN tumour IS malignant

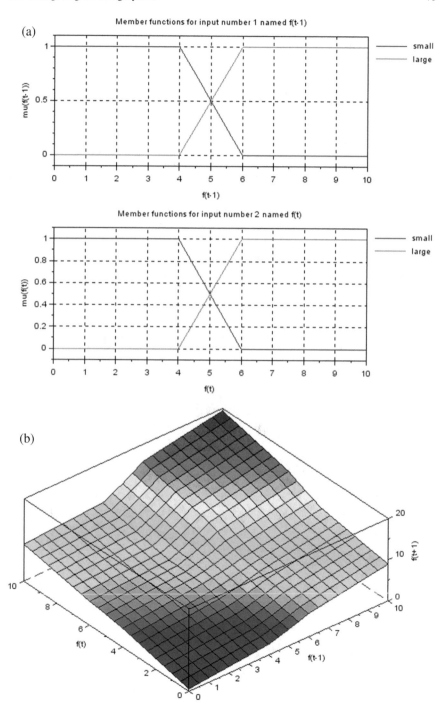

Fig. 5.8 TSK FIS example: **a** two input variables **b** 3D-plot of the output

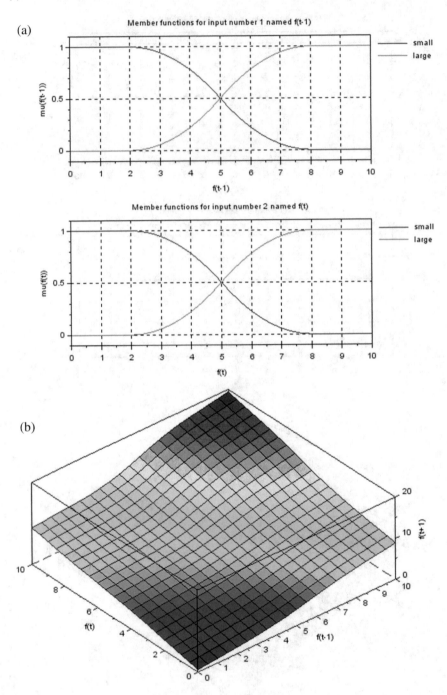

Fig. 5.9 TSK FIS example: **a** alternative input variables **b** 3D-plot of the output

Table 5.1 Input features for breast cancer classification

Feature	Abbrev.	UoD
Clump Thickness	CT	[1,10]
Uniformity of Cell Size	UCZ	[1,10]
Uniformity of Cell Shape	UCS	[1,10]
Marginal Adhesion	MA	[1,10]
Single Epithelial Cell Size	SECS	[1,10]
Bare Nuclei	BN	[1,10]
Bland Chromatin	BC	[1,10]
Normal Nucleoli	NN	[1,10]
Mitoses	MS	[1,10]

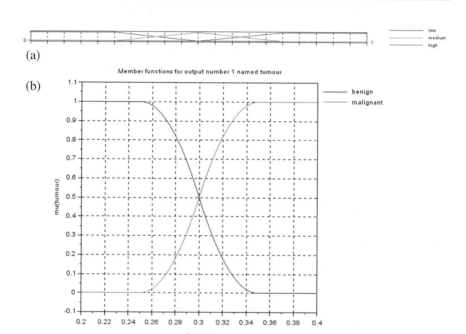

Fig. 5.10 Mamdani FIS for breast cancer classification: **a** terms of any input variable **b** terms of the output variable

R5: IF CT IS high AND UCZ IS high AND UCS IS high AND MA IS medium AND SECS IS medium AND BN IS high AND BC IS medium AND NN IS low AND MS IS low THEN tumour IS malignant

R6: IF CT IS medium AND UCZ IS low AND UCS IS low AND MA IS low AND SECS IS low AND BN IS low AND BC IS low AND NN IS low AND MS IS low THEN tumour IS benign

R7: IF CT IS high AND UCZ IS high AND UCS IS medium AND MA IS high AND SECS IS high AND BN IS high AND BC IS medium AND NN IS medium AND MS IS medium THEN tumour IS malignant

R8: IF CT IS low AND UCZ IS low AND UCS IS low AND MA IS medium AND SECS IS low AND BN IS low AND BC IS low AND NN IS low AND MS IS low THEN tumour IS benign

R9: IF CT IS medium AND UCZ IS medium AND UCS IS medium AND MA IS medium AND SECS IS medium AND BN IS medium AND BC IS medium AND NN IS medium AND MS IS low THEN tumour IS benign

R10: IF CT IS high AND UCZ IS high AND UCS IS high AND MA IS high AND SECS IS high AND BN IS high AND BC IS high AND NN IS high AND MS IS low THEN tumour IS malignant

R11: IF CT IS low AND UCZ IS medium AND UCS IS low AND MA IS low AND SECS IS low AND BN IS low AND BC IS low AND NN IS low AND mitoses IS low THEN tumour IS malignant

R12: IF CT IS low AND UCZ IS low AND UCS IS medium AND MA IS low AND SECS IS low AND BN IS low AND BC IS low AND NN IS low AND mitoses IS low THEN tumour IS benign

R13: IF CT IS medium AND UCZ IS medium AND UCS IS medium AND MA IS high AND SECS IS medium AND BN IS medium AND BC IS medium AND NN IS medium AND MS IS low THEN tumour IS malignant

The performance measures are given below:

- Accuracy = 0.9764706,
- Precision = 0.6666667,
- Recall = 0.6666667,
- F-measure = 0.6666667.

These results are obtained in Scilab by the following instructions, after variables initialization and file reading, forming the input matrix X and the target vector T

```
fls=loadfls("breastcc.fls");
for i=1:length(T)
y=evalfls(X(i,:),fls);
if T(i)<3 & y<3
then TN=TN+1;
else if T(i)>=3 & y<3
then FN=FN+1;
else if T(i)<3 & y>=3
then FP=FP+1;
else TP=TP+1;
end
end
end
end
Accuracy=(TP+TN)/(TP+TN+FP+FN)
```

```
Precision = TP / (TP+FP)
Recall = TP / (TP+FN)
Fmeasure = 2*(Precision * Recall) / (Precision + Recall)
```

where `TN`, `TP`, `FP`, `FN` are the variables true negative, true positive, false positive, false negative, respectively. The instruction `loadfls` allows loading the .fls file generated by the Scilab Fuzzy Logic Toolbox, while `evalfls` is the instruction to use the .fls structure in inference mode to get the output given some input data. The `disp` command can be used to show the results.

It is worth mentioning that this data set has already been considered in literature for classification by a Mamdani system. In a recent paper (Hernandez-Julio et al. 2019), the authors reduce the number of input variables to 5 by means of clustering, but each variable has 10 terms (whose linguistic name is not mentioned) and there are 192 rules. The accuracy they found is 99.28% and F-measure 0.9897%.

Problems

1. Calculate the center of maximum of the generic triangular and trapezoidal fuzzy numbers, $A = (a_1, a_2, a_3)$ and $B = (b_1, b_2, b_3, b_4)$ respectively.
2. Compute the centroid defuzzifier for $A = (-3, 0, 4)$.
3. Compute the centroid defuzzifier for $A = \{(-2, 0.8), (-1, 0.5), (0, 0.3), (1, 1), (2, 0.4)\}$.
4. Consider a single input-single output TSK FIS. The aim is to approximate the function $2Cos(x/2)$ with $x \in [0, \pi]$. Elaborate the fuzzy partition in the input variable space and the number of required rules.

References

Hernandez-Julio YF et al (2019) Framework for the development of data-driven mamdani-type fuzzy clinical decision support systems. Diagnostics 9:52

Jang J-SR, Sun C-T, Mizutani E (1997) Neuro-fuzzy and soft computing. Prentice Hall, Upper Saddle River

E H Mamdani, S Assilian (1975) An experiment in linguistic synthesis with a fuzzy logic controller, Int J Man-Machine Studies, 7(1): 1–13

Takagi T, Sugeno M (1985) Fuzzy identification of systems and its applications to modeling and control. IEEE Trans Syst Man Cybern 15:116–32

Zadeh LA (1975) The concept of linguistic variables and its application to approximate reasoning I, II, III. Inf Sci 8:199–249, 301–357, 43–80

Chapter 6
Combining Artificial Neural Networks and Fuzzy Sets

6.1 A Brief Introduction to Artificial Neural Networks

The Artificial Neural Network (ANN) is a computing scheme providing a mapping between input and output variables. The input variables are also called attributes or features. ANNs are employed both in regression and classification problems.

The basic component of the ANN is the *neuron* (also called unit). An ANN consists of several layers equipped with a different number of neurons, namely, input, hidden, and output layers. There may be one or more hidden layers (Fig. 6.1). The input ξ of a generic node k is given by the linear sum of incoming signals x_i from the preceding layer, through the weights w_{ik}, plus a bias term b_k, that is

$$\xi_k = \sum_i w_{ik} x_i + b_k. \tag{6.1}$$

The generic incoming signal is given by $x_i = f(\xi_i)$, where f is the activation function. The weights are arranged into matrices.

There are different activation functions, e.g.

- *linear:* $f(\xi_i) = c\xi_i$, that is the identity function when $c = 1$; it is mostly used in the output node;
- *sigmoid:* $f(\xi_i) = \frac{1}{1+\exp(-\xi_i)}$;
- *hyperbolic tangent:* $f(\xi_i) = \frac{2}{1+\exp(-2\xi_i)} - 1$;
- *Rectified Linear Unit (ReLU):* $f(\xi_i) = max\{0, \xi_i\}$, which has the advantage of a lower processing time.

The number of neurons and their distribution into layers describe the architecture of an ANN. This is usually fixed by the user, according to the experience. It is worth recalling two main types of architectures:

- feedforward, where the model has connections only in one direction from the input to the output layer, without any connection among the neurons of the same layer;

© The Author(s), under exclusive license to Springer Nature Switzerland AG 2022
S. Tomasiello et al., *Contemporary Fuzzy Logic*, Big and Integrated
Artificial Intelligence 1, https://doi.org/10.1007/978-3-030-98974-3_6

Fig. 6.1 One-hidden layer
ANN

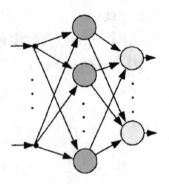

- feedback, the output from a layer can be directed back to the same layer or any other layer; recurrent neural networks (RNNs) belong to this class.

Learning occurs in a brain because of the formation and modification of synapses. In the computing scheme, the learning is the act of optimizing a mathematical model. In general, learning rules can be classified into three categories, i.e. supervised, unsupervised and combination of the two.

In supervised learning mode, an ANN is supplied with a training data set

$$D = \{\mathbf{x}^{(k)}, \overline{\mathbf{y}}^{(k)}\}_{k=1}^{N} = \{(x_1^{(k)}, \ldots, x_m^{(k)}), (\overline{y}_1^{(k)}, \ldots, \overline{y}_q^{(k)})\}_{k=1}^{N}, \quad (6.2)$$

consisting of N input-output data samples. When each input $\mathbf{x}^{(k)}$ is presented to the network, the corresponding desired output $\overline{\mathbf{y}}^{(k)}$ is also supplied. The difference between the computed output $\mathbf{y}^{(k)} = (y_1^{(k)}, \ldots, y_q^{(k)})$ and the desired output $\overline{\mathbf{y}}^{(k)}$ is calculated in order to correct the weights so that the actual output will move closer to the desired output.

In unsupervised learning mode, the desired outputs are not available. Therefore, there is no feedback to assess whether the outputs are correct. The learning algorithm must discover, on its own, patterns, features, regularities, correlations, or categories in the input data and code for them in the output.

As mentioned before, the supervised training of a neural network can be regarded as a problem in numerical optimization. In general, the optimization is about minimizing or maximizing an objective function. In an ANN, this function (error function) depends on the model's parameters, i.e. weights and biases. The optimization may involve only first-order derivatives or also second-order derivatives. In the first case, the algorithm minimizes the objective function using gradient values with respect to the parameters (e.g. gradient descent algorithm). In the second case, the second-order derivative values with respect to the parameters are used to minimize the objective function (e.g. conjugate-gradient methods, Levenberg–Marquardt algorithm).

A typical supervised learning algorithm is the *backpropagation*. It is a specific iterative technique for implementing the gradient descent in the weights space. Weights at the nth iteration (epoch) are updated by the *delta rule*:

$$w_{ij}(n) = w_{ij}(n-1) + \Delta w_{ij}(n-1), \tag{6.3}$$

with $\Delta w_{ij}(n-1) = -\eta \frac{\partial E(n-1)}{\partial w_{ij}(n-1)}$ where $E(n-1) = \frac{1}{2}\sum_j(\overline{y}_j - y_j(n-1))^2$, with $\overline{y}_j$ being the jth target value.

By using the chain rule of calculus, $\Delta w_{ij}(n-1)$ can be written as

$$\Delta w_{ij}(n-1) = -\eta \frac{\partial E(n-1)}{\partial y_j(n-1)} \frac{\partial y_j(n-1)}{\partial \xi_j(n-1)} \frac{\partial \xi_j(n-1)}{\partial w_{ij}(n-1)} = \eta \delta_j x_i(n-1)$$

where $\delta_j = [\overline{y}_j - y_j(n-1)]\frac{\partial f(\xi_j)}{\partial \xi_j}$ is the error signal of the jth node in the output layer for the kth data sample.

Hence, in the network there are two phases:

- forward phase, where the inputs of a training instance are fed into the neural network; the computed output is compared to the target of the training instance and the gradient values of the loss function (with respect to the weights) are computed;
- backward phase, where the gradient of the loss function is learned to update the weights.

The backpropagation algorithm is considered to have converged when the absolute rate of change of the average squared error per epoch is small enough, e.g. it is in the range [0.001, 0.01].

Important factors determining the convergence of the backpropagation learning algorithm are the initial weights and the learning rate η. The initial weights of a multilayer feedforward network are typically small random values, in order to prevent a local minimum. The learning rate represents the rate at which the network's weights are updated. The lower the learning rate, the smaller the changes in the weights from one iteration to the next, the higher the computational time.

Once the network has been trained, there is a testing stage, where unseen data are presented to the network.

The problem of overfitting refers to the fact that fitting a model to a particular training data set does not guarantee its good prediction performance on unseen test data. There is always a gap between the training and test data performance, which may be significant when the model is complex and the data set is small.

It is possible to combine the artificial neural networks (ANNs) and the theory of fuzzy sets and fuzzy logic mainly in two ways:

- by using the theory of fuzzy sets and fuzzy logic within the framework of the ANNs, e.g. as a generalization of the classical ANNs to process quantitative (numerical) data and qualitative (linguistic) information represented by means of fuzzy sets (fuzzy neural networks);

- by using ANNs within the framework of fuzzy modelling and design of fuzzy systems, for the automatic tuning of their parameters (neuro-fuzzy systems).

6.2 A Fuzzy Neural Network

Several types of fuzzy neural networks have been proposed. Here we recall a general scheme to emphasize the difference with respect to a neuro-fuzzy system.

Consider n elements $x_1, x_2, \ldots, x_n$, with $x_i \in X_i \subset \mathbb{R}$, $i = 1, 2, \ldots, n$ and m elements $y_1, y_2, \ldots, y_m$, with $y_j \in Y_j \subset \mathbb{R}$, $j = 1, 2, \ldots, m$. The training data for a fuzzy neural network has the form of K input-output pairs, that is,

$$\mathbf{D} = \{\overline{\mathbf{A}}_k, \overline{\mathbf{B}}_k\}_{k=1}^K \tag{6.4}$$

where $\overline{\mathbf{A}}_k = \{\overline{A}_{1k}, \overline{A}_{2k}, \ldots, \overline{A}_{nk}\}$ and $\overline{\mathbf{B}}_k = \{\overline{B}_{1k}, \overline{B}_{2k}, \ldots, \overline{B}_{nk}\}$. $\overline{A}_{ik}$ and $\overline{B}_{jk}$ can be linguistic terms and numerical data.

The linguistic terms are represented by fuzzy sets, i.e. $\overline{A}_{ik} \in \mathcal{F}(X_i)$ and $\overline{B}_{jk} \in \mathcal{F}(Y_j)$, where $\mathcal{F}(X_i)$ and $\mathcal{F}(Y_j)$ denote the families of all the fuzzy sets defined in the universes X_i and Y_j, respectively. The numerical data are represented by means of fuzzy singletons.

The fuzzy neural network consists of a conventional neural network and two interfaces, one for the input and one for the output, built on the basis of the fuzzy sets theory. The network is employed first in a learning mode, to train it, and then in inference mode, to provide an approximate solution for the considered problem.

The input and output interfaces are represented by means of a collection of fuzzy sets, separately for each system input x_i, $i = 1, 2, \ldots, n$, and each output y_j, $j = 1, 2, \ldots, m$. The system may process both numerical data and fuzzy data. The collection of fuzzy sets represents a fuzzy partition of a given input or output domain.

Let $A = \{A_1, A_2, \ldots, A_n\}$ be a sequence of continuous normal fuzzy sets A_i : $X \to [0, 1]$. A is a partition of X, if $\sum_{i=1}^n A_i(x) = 1$, $\forall x \in X$. This requirement could be made less restrictive by requesting that each element in X belongs to at least one A_i with nonzero membership level, that is $\forall x \exists i$ s.t. $A_i(x) > 0$. Hence, the main requirement for the fuzzy sets is that each element $x_i \in X_i$ (or $y_j \in Y_j$) must belong to at least one fuzzy set with a strictly positive degree of membership. The fuzzy sets are usually defined by an expert. For instance, in the medical field, many parameters are characterized by three linguistic terms: "normal", "high" and "low".

Let us assume that for each input x_i, $i = 1, 2, \ldots, n$, a collection $\{A_{i1}, A_{i2}, \ldots, A_{ia_i}\}$ of a_i fuzzy sets has been defined. Analogously, for each y_j, $j = 1, 2, \ldots, m$, a collection $\{B_{j1}, \ldots, B_{j2}, \ldots, B_{jb_j}\}$ of b_j fuzzy sets has been fixed.

The transformed input data has the form of a set of activation degrees χ of the fixed fuzzy sets for a given input

$$\chi(A'_i/A_{il_i}) = \sup_{x_i}\{min[\mu_{A'_i(x_i)}, \mu_{A_{il_i}(x_i)}]\}. \tag{6.5}$$

Analogously, for the output there is a set of desired activation degrees η of the particular fuzzy sets for a given output:

$$\eta(B'_j/B_{jl_j}) = \sup_{y_j}\{min[\mu_{B'_j(y_j)}, \mu_{B_{jl_j}(y_j)}]\} \tag{6.6}$$

$j = 1, \ldots, m, l_j = 1, \ldots, b_j$.

In particular, for nonfuzzy numerical data $x'_i \in X_i$, the input fuzzy set A'_i is reduced to the fuzzy singleton $\overline{x}'_i$ described by the membership function

$$\mu_{\overline{x}'_i} = \begin{cases} 1, & \text{for } x_i = x'_i \\ 0, & \text{for } x_i \neq x'_i \end{cases} \tag{6.7}$$

and the activation degree becomes

$$\chi(x'_i/A_{il_i}) = \sup_{x_i \in X_i}\{min[\mu_{\overline{x}'_i}, \mu_{A_{il_i}(x_i)}]\} = \mu_{A_{il_i}(x'_i)}. \tag{6.8}$$

The generated input activation degrees are then processed by a conventional neural network, which generates the activation degrees for the outputs. The latter are, compared with the corresponding desired activation degrees. The differences between the latter and the computed activation degrees for outputs are then processed by a learning algorithm, which adjusts the weights of the conventional neural network in order to minimize these differences. The cost function, which is minimized during the learning process, clearly depends on the difference between desired activation degrees and the computed activation degrees for the outputs. Backpropagation is used to adjust the weights of the neural network, reducing the cost function to an acceptable value. Once the learning phase is successfully completed, the fuzzy neural network can be employed as an approximate inference and forecasting tool.

The fuzzy neural network in inference mode works as follows. The input data are first processed by the input interface and transformed in the form of activation degrees. These ones are presented to the trained neural network which produces the activation degrees for the outputs. The latter are passed to the output block, which produces the output fuzzy sets and provides their defuzzification, when nonfuzzy numerical responses are required.

6.3 Adaptive Neuro-Fuzzy Inference System (ANFIS)

One of the first neuro-fuzzy systems for rule-based function approximation was the Adaptive Neuro-Fuzzy Inference System (ANFIS) introduced by Jang (1993). It represents the first-order Takagi-Sugeno (TS) fuzzy inference system in a five-layer network architecture, as detailed below.

Layer 1 (L1). It consists of adaptive nodes i whose output is the membership degree of the input x to the fuzzy set A_{ir}

$$O_i = \mu_{A_{ir}}(x), \quad i = 1, 2... \tag{6.9}$$

where x is the input to node i and A_{ir} is a linguistic term associated with this node. The membership function for A_{ir} can be any suitable parameterized membership function such as the generalized bell-shaped function:

$$\mu_{A_i}(x) = \left(1 + |\frac{x - c_i}{a_i}|^{2b_i}\right)^{-1}, \tag{6.10}$$

where $\{a_i, b_i, c_i\}$ is the parameter set. These parameters, whose values change, are referred to as *premise parameters*.

Layer 2 (L2). It consists of fixed nodes, whose output represents the firing strength of a rule, usually computed by means of product-type t-norms (to model AND)

$$w_r = \Pi_{i=1}^n \mu_{A_{ir}}(x_i) \tag{6.11}$$

Layer 3 (L3). Every node in this layer is a fixed node. The outputs of this layer are the *normalized firing strengths*, i.e. the i-th node calculates the ratio of the i-th rule's firing strength to the sum of all rules' firing strengths:

$$\overline{w}_r = w_r / \sum_{j=1}^{R} w_j. \tag{6.12}$$

Layer 4 (L4). This layer's nodes are adaptive, with node function

$$o_r = \overline{w}_r y_r = \overline{w}_r \theta_{0r} + \theta_{1r} x_1 + \cdots + \theta_{nr} x_n, \tag{6.13}$$

where $\overline{w}_r$ is a normalized firing strength from L3 and $\{\theta_{0r}, \theta_{1r}, \ldots, \theta_{nr}\}$ is the parameter set of this node. These parameters are referred to as *consequent parameters*.

Layer 5 (L5). The single (fixed) node computes the final output by summing all the outputs from L_4

$$y^0 = \sum_{r=1}^{R} o_r. \tag{6.14}$$

In general, ANFIS implements rules of the form

$$\text{IF } x_1 \text{ is } A_{1r} \text{ AND }...\text{AND } x_n \text{ is } A_{nr}$$

$$\text{THEN} \quad y_r = \theta_{0r} + \theta_{1r}x_1 + \cdots + \theta_{nr}x_n.$$

ANFIS aims at adjusting the antecedents and the consequent parameters, by using a hybrid learning algorithm based both on backpropagation and least-squares (LS) method. The node outputs go forward until L4 and the consequent parameters are identified by the LS method for fixed antecedent parameters. This means that, by using the training data, one obtains a matrix equation such as $\mathbf{H}\theta = \mathbf{y}$, where θ collects the unknown parameters and $\mathbf{y}$ the target values. The best solution for θ which minimizes $\|\mathbf{H}\theta - \mathbf{y}\|^2$ is $\theta* = (\mathbf{H}^T\mathbf{H})^{-1}\mathbf{H}^T\mathbf{y}$, where $(\mathbf{H}^T\mathbf{H})^{-1}\mathbf{H}^T$ is the pseudoinverse of $\mathbf{H}$. Backpropagation is then used to adjust the antecedent parameters.

Each epoch in the learning procedure is composed of two parts: forward pass and backward pass. In the first part, the input patterns are propagated, and the optimal consequent parameters are estimated by a least-squares procedure, while the antecedent parameters are assumed to be fixed. In the second part, the error signals propagate from the last layer toward the first layer. In this part, backpropagation is used to modify the antecedent parameters, while the consequent parameters remain fixed. The procedure is then iterated.

ANFIS is related to Multiple Input Single Output (MISO) systems. It provides the automatic tuning of a Sugeno-type inference system, generating a single output of a weighted linear combination of the consequents. Its generalization to Multiple Input Multiple Output (MIMO) systems with nonlinear fuzzy rules is CANFIS, which stands for Coactive neuro-fuzzy inference system Jang et al. (1999). Here the final ith output, in presence of r rules and n inputs is computed as $y_i = f_i(\sum_{j=1}^{r} \overline{w}_j C_{ji})$, with $C_{ji} = g_i(\sum_{k=1}^{n} p_{jk}x_k + r_j)$, where f_i and g_i can be in general nonlinear functions.

6.4 An ANFIS Variant

The standard ANFIS uses grid partitioning to generate fuzzy rules. This method divides the input space into a grid-like form, and each grid partition represents a fuzzy rule. This implies that that the number of rules increases exponentially as the number of input variables increases. An alternative is represented by the scatter partitioning, where the input space is usually divided into an arbitrary number of clusters. The most used clustering methods in this context are the fuzzy C-means (FCM) and subtractive clustering, even though they may have a certain computational cost (Yeom and Kwak 2018). It is clear that the grid-based and clustering-based schemes both present limitations when applied to mid- or high-dimensional regression problems. The standard ANFIS uses a hybrid learning algorithm based both on backpropagation and least-squares (LS) method. It is well known that backpropagation, being a gradient based technique, may incur local minima. This disadvantage can be weakened by using global-optimization techniques. This is the reason why some authors revised the ANFIS' learning algorithm by introducing population based algorithms, such as genetic algorithm (GA), ant colony optimization (ACO), particle swarm optimization (PSO), artificial bee colony (ABC) (Shihabudheen and Pillai 2018). Anyway, such

techniques may have a significant computational cost. In order to lower the computational cost, it is possible to use only the LS method (Pramod and Pillai 2021), but
even in this case there may be drawbacks. In fact, the computation of the generalized
inverse matrix in LS methods can be cumbersome or not feasible for ill-conditioned
problems. The regularization techniques aim at addressing such issues.

Here we focus on the variant with fractional regularization. In this variant, the
learning algorithm is based only on the Tikhonov regularization. The fractional
Tikhonov regularization represents a generalization of the standard Tikhonov method,
which is usually employed for solving discrete ill-posed inverse problems. The minimization problem for the standard method is formulated as follows:

$$\min_{\beta} \|\mathbf{H}\theta - \mathbf{y}\|^2 + \lambda\|\theta\|^2, \tag{6.15}$$

where θ collects the unknown consequents parameters, as mentioned before, and
$\lambda \in \mathbb{R}^+$ is a regularization parameter. The solution is:

$$\theta* = (\mathbf{H}^T\mathbf{H} + \lambda\mathbf{I})^{-1}\mathbf{H}^T\mathbf{y}, \tag{6.16}$$

where $\mathbf{I}$ is the identity matrix with dimension nR.

The fractional Tikhonov method formalizes the following minimization problem

$$\min_{\theta} \|\mathbf{H}\theta - \mathbf{y}\|_P^2 + \lambda\|\theta\|^2, \tag{6.17}$$

where $\|\theta\|_P = (\theta^T\mathbf{P}\theta)^{\frac{1}{2}}$ and $\mathbf{P}$ is a symmetric positive semi-definite matrix defined
as (Hochstenbach and Reichel 2011)

$$\mathbf{P} = (\mathbf{H}^T\mathbf{H})^{\frac{\alpha-1}{2}}. \tag{6.18}$$

The solution can be written as:

$$\theta* = (\mathbf{M}^{\frac{\alpha+1}{2}} + \lambda\mathbf{I})^{-1}\mathbf{M}^{\frac{\alpha-1}{2}}\mathbf{H}^T\mathbf{y}, \tag{6.19}$$

where $\mathbf{M} = \mathbf{H}^T\mathbf{H}$. It is clear that when $\alpha = 1$ then (6.19) reduces to (6.16), that is the
standard regularization. The case without any regularization is obtained by setting
$\lambda = 0$ and $\alpha=1$. The choice of λ and α is critical to the accuracy.

The aim of using such learning algorithm is to ensure good accurary and low computational cost, avoiding backpropagation and grid partitioning. By grid partitioning, the rules are generated by enumerating all possible combinations of membership
functions of all inputs, making the number of fuzzy rules increase exponentially with
the number of input variables.

In the ANFIS variant here presented, named ANFIS-T (Tomasiello et al. 2022),
the number of rules R equals the number of terms m for each variable. For each

Fig. 6.2 ANFIS scheme

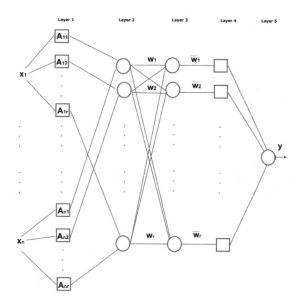

variable, the membership functions (e.g. Gaussian) are fixed to form a uniform fuzzy partition of the variable domain. The jth rule is

$$\text{IF } x_1 \text{ is } A_{1j} \ldots \text{ AND } x_n \text{ is } A_{nj}$$

$$\text{THEN } y_j = \theta_{0j} + \theta_{1j}x_1 + \cdots + \theta_{nj}x_n,$$

with $j = 1, \ldots, R$. Figure 6.2 depicts the ANFIS scheme for the case of two input variables (for the sake of simplicity), with number of rules equal to the number of terms.

The variant's scheme has generally a low number of rules and terms, implying a low complexity. This results in a lower computational cost, and it is also beneficial to interpretability. Besides, the distinguishability constraint is not violated, because the membership functions are pre-defined for each input variable and remain fixed during the learning process, by retaining their well-defined semantics.

6.4.1 The ANFIS Variant Code

The Scilab function `[Y,rmse2]=anfist(In,Int,T,Tt,NumInTerms, C,a)` implements the presented ANFIS variant. The function returns the computed test output values and the test RMSE. The function arguments are:

- `In`, input data matrix for training (number of training instances × number of attributes);

- T, target vector for training;
- Int, input data matrix for testing (number of testing instances × number of attributes);
- Tt, target vector for testing;
- NumInTerms, number of terms for each variable;
- C, standard regularization parameter;
- a, fractional regularization parameter (when a=1 the standard regularization occurs).

In this variant it is assumed that the number of rules equals the number of terms for each input variable, i.e. NumRules = NumInTerms. In this code, the Gaussian MF has been used, but the option to choose the MF could be easily programmed.

In the training stage, the outputs Out1, Out2, Out3 of the layers L1, L2, L3 are first computed:

```
for j=1:NumSamples
In2 = In0(:,j)*ones(1,NumInTerms);
Out1 = exp(-((In2-alpha)./sigma).^2);
Out2 = prod(Out1,1);
S_2 = sum(Out2);
if S_2 ~= 0
Out3 = Out2 ./S_2;
else
Out3 = zeros(1,NumRules);
end
```

where NumSamples=size(In,1); is the number of training samples (i.e. rows of the input data matrix for training). Then the matrix H is formed and the vector Theta of the unknown consequent parameters is computed.

```
for k=1:NumRules
H0{j,k}=Out3(k)*In'(j,:);
end
H=cell2mat(H0);
M=H'*H;
PH=inv(eye(M)*C+M^((a+1)/2))*M^((a-1)/2)*H';
Theta=PH*T1
```

In the test stage, the outputs of the different layers are computed likewise, although using the data for testing. The main difference is the final part: the vector Theta, previously determined, is here used to compute the final output, once the matrix H2 (related to the test input data) has been formed:

```
H2=cell2mat(H0t);
Y=H2*Theta;
```
Finally, the RMSE is computed
```
ERt=T1t-Y;
rmse2= norm(ERt)/sqrt(NumSamplest);
```

Example 6.1 Predicting chaotic time series

The time series here considered is generated by the chaotic Mackey–Glass (MG) time-delay differential equation

$$\frac{dx}{dt} = \frac{\gamma x(t - \tau)}{1 + x^\eta(t - \tau)} + \beta x, \tag{6.20}$$

where $\eta = 10, \gamma = 0.2, \beta = -0.1$. When $\tau > 16.8$, the MG system exhibits a chaotic behaviour. Thus $\tau = 17$ is chosen.

The aim is to predict the value $x(t + 6)$ by considering the four input values $x(t - 18), x(t - 12), x(t - 6), x(t)$.

By solving numerically (6.20), 1000 data samples, between $t = 118$ and 1117, have been created. The first 500 data samples were used as the training data set and the remaining 500 samples as the test data set.

Three terms for each one of the four input variables have been fixed. Their linguistic meaning is *small, medium, large*. The best result by ANFIS-T was sought by choosing $\alpha = 0.95$ and $\lambda = 0.01$. The root mean square error (RMSE) was $RMSE = 9.85E - 02$. It is worth mentioning that only 3 rules were used and the training time was only 0.06 s (Intel i5, 10th generation processor). By ANFIS with a reduced 20-rule base, the error is similar or slightly lesser $RMSE = 9.315E - 02$, as reported by Gorzalczany (2002). Anyway, at least 100 epochs are needed to achieve such a result by ANFIS, implying a certain computational cost.

The numerical experiments in Scilab were run by the following instructions, whose meaning is illustrated in comment mode:

```
exec('anfist.sce');//execute the required function
SS=readxls('data.xls');//read file xls
(in the current directory)
S1=SS(1);//first worksheet with training data
S2=SS(2);//second worksheet with test data
X1=S1(:,1:size(S1,2)-1);//input matrix (training)
T1=S1(:,size(S1,2));//target values (training)
X2=S2(:,1:size(S1,2)-1);//input matrix (testing)
T2=S2(:,size(S1,2));//target values (testing)
C=0.1;
a=0.1;
NumInTerms=3;
tic()
[Y, rmse]=anfisf(X1,X2,T1,T2,NumInTerms,C,a);
tt=toc();
disp('RMSE:',rmse,'time:',tt,'terms:',NumInTerms,
'C:',C,'a:',a)
```

Example 6.2 Predicting carbon dioxide emissions in an industrial gas furnace system (Box-Jenkins data set)

The considered time series consists of observations pairs (Box et al. 2009): the

Table 6.1 Box-Jenkins data set: comparison against the RBF results by Harpham and Dawson (2006), first testing data set

Approach	MSE
RBF (multiquadratic)	0.145
RBF (inverse multiquadratic)	0.116
RBF (Gaussian)	0.153
ANFIS-T (4 terms, $\lambda = 0.001$, $\alpha = 0.9$)	0.135

Table 6.2 Box-Jenkins data set: comparison against the RBF results by Harpham and Dawson (2006), second testing data set

Approach	MSE
RBF (multiquadratic)	0.070
RBF (inverse multiquadratic)	0.050
RBF (Gaussian)	0.099
ANFIS-T (3 terms, $\lambda = 0.001$, $\alpha = 0.9$)	0.0841

methane gas feed rate ($u(t) \in U$) and the concentration of CO_2 in the exhaust gases ($y(t) \in Y$).

The aim is to predict $y(t)$ based on $\{y(t-1), \ldots, y(t-4), u(t-1), \ldots, u(t-6)\}$. The number of data samples is 290, of which 145 are used for training and 145 for testing. By following Harpham and Dawson (2006) for comparative purposes, the two data sets consisting of 145 samples were alternately used for training and testing. The above-mentioned authors considered radial basis function (RBF) networks, with different kinds of basis functions. As pointed out by Jang et al. (1999), ANFIS without backpropagation can be regarded as a kind of RBF network. Some results by ANFIS-T are tabled in Tables 6.1 and 6.2 for the two testing sets. The other results appearing in the same tables are by the above-mentioned authors. In both cases, the Gaussian membership function was used for the terms of each input variable. It can be observed that the performance of ANFIS-T in terms of mean squared error (MSE) is better than RBF network with Gaussian basis functions, but slightly worse than the RBF network with inverse multiquadratic. It is worth mentioning that by means of ANFIS-T, with 3 terms, $\lambda = 0.1$ and $\alpha = 0.3$, by using 10-fold cross validation, the average MSE is $MSE = 0.059$, which is lesser than the reported average MSE by the RBF network with inverse multiquadratic.

The Scilab instructions to perform the numerical experiments are similar to the ones reported in the previous example.

Example 6.3 Predicting the total cost to serve in a supply chain

In this example, the results by ANFIS-T are compared against the ones by the traditional ANFIS with grid partitioning and ANFIS with fuzzy C-means (FCM) as reported by Tomasiello et al. (2022).

The problem is to predict the total cost to serve in a supply chain (SC), given some variables as detailed in Table 6.3. Such variables represents the level-2 metrics

Table 6.3 Input variables to predict the total cost to serve

Variable	UoD ($\times 10^3$)	MU
Sourcing cost	[140, 300]	USD
Planning cost	[25, 50]	USD
Material landed cost	[70, 150]	USD
Production cost	[150, 380]	USD
Order management cost	[220, 480]	USD
Fulfillment cost	[45, 70]	USD
Returns cost	[50, 200]	USD
Cost of goods sold	[1300, 1900]	USD

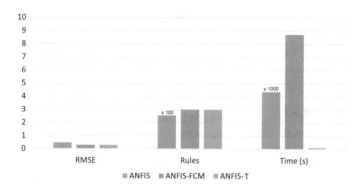

Fig. 6.3 Results for the cost to serve in a SC

of the SCOR model, which is a well-known scheme for supply chain performance evaluation. The total cost to serve is a level-1 metric. Its value may vary in the range [2,000,000, 3,530,000] USD. The universe of discourse (UoD) and measurement unit (MU) of the input values appear in Table 6.3.

The values to perform the experiments were uniformly random generated. There were 1000 random data points generated for each variable. The data was normalized using the min-max normalization. The data set was split into two parts, with 70% of the samples used for the training and 30% for testing. Figure 6.3 shows RMSE, number of rules and training time for the three approaches. The number of rules for ANFIS is 2^8 and the training time is greater than 4000 s. ANFIS-FCM and ANFIS-T ($\lambda = 0.01$, $\alpha = 0.9$) with 3 rules have the same RMSE, but the training time for ANFIS-FCM is almost 90 times greater.

Table 6.4 Classification of breast cancer: performance of ANFIS-T for different values of the parameters α and λ

Approach	Accuracy	Precision	Recall	F-measure	Time (s)
ANFIS-T (2 terms, $\lambda = 0, \alpha = 1$)	0.9729	0.973	1	0.986	0.048
ANFIS-T (3 terms, $\lambda = 0, \alpha = 1$)	0.9787	0.9787	1	0.9892	0.061
ANFIS-T (2 terms, $\lambda = 1, \alpha = 0.9$)	0.9911	0.9687	1	0.9841	0.051
ANFIS-T (3 terms, $\lambda = 1, \alpha = 0.9$)	0.975	0.975	1	0.987	0.087

Example 6.4 Classification of breast cancer

The aim is to classify whether a tumour is malignant or not, We refer to the same data set and input features described in Example 5.7. The performance by ANFIS-T for different values of the parameters α and λ is shown in Table 6.4. The best one is by ANFIS-T with 2 terms, $\lambda = 1, \alpha = 0.9$. The related tabled values are better than the ones discussed in Example 5.7 by the classical Mamdani FIS and the one using clustering.

Problems

1. Implement in Scilab the input and output interfaces to get a fuzzy neural network.
2. Implement different kinds of parametric membership functions.
3. Consider a single input ANFIS-T scheme. The aim is to approximate the function $2Cos(x/2)$ with $x \in [0, \pi]$. Discuss the solution and differences against the one sought in Problem 4, Chap. 5.

References

Box GEP, Jenkins GM, Reinsel GC (2009) Time series analysis, forecasting and control. Wiley, Hoboken

Gorzalczany MB (2002) Computational intelligence systems and applications. Springer, Berlin

Harpham C, Dawson C W , (2006) The effect of different basis functions on a radial basis function network for time series prediction: A comparative study, Neurocomputing. 69: 2161–2170

Hochstenbach ME, Reichel L (2011) Fractional Tikhonov regularization for linear discrete ill-posed problems. BIT Num Math 51:197–215

Jang JSR (1993) ANFIS: adaptive-network-based fuzzy inference system. IEEE Trans Syst Man Cyber 23(5/6):665–685

Jang J-SR, Sun C-T, Mizutani E (1999) Neuro-fuzzy and soft computing. Prentice Hall

Pramod CP, Pillai GN (2021) K-means clustering based extreme learning anfis with improved interpretability for regression problems. Know Bas Syst 215:106750

Shihabudheen KV, Pillai GN (2018) Recent advances in neuro-fuzzy system: a survey. Know Bas Syst 152:136–162

Tomasiello S, Pedrycz W, Loia V, (2022) On Fractional Tikhonov Regularization: Application to the Adaptive Network-Based Fuzzy Inference System for Regression Problems. IEEE Transactions on Fuzzy Systems, https://doi.org/10.1109/TFUZZ.2022.3157947

Tomasiello S, Uzair M, Loit E (2022) ANFIS with fractional regularization for supply chains cost and return evaluation. In: Proceedings of 13th international workshop on fuzzy logic and applications (WILF2021), p 70

Yeom C, Kwak K-C (2018) A performance comparison of ANFIS models by scattering partitioning methods. In: Proceedings of 2018 IEEE 9th information technology, electronics and mobile communication conference (IEMCON), pp 814–818

Chapter 7
Fuzzy Transform

7.1 Direct and Inverse Fuzzy Transform

The fuzzy transform (F-transform or FT for short) is an approximation technique based on the fuzzy set theory which was introduced by Perfilieva (2006). Like any well–known transform, such as Laplace, Fourier, it consists of direct and inverse phases. Unlike other transforms, it uses a fuzzy partition of a universe. It is based on a linear combination of membership functions, here called basic functions. The F–transform provides a linear mapping from a space with a certain dimension to a space of lower dimension, The inverse F-transform returns to the original space, offering the approximate solution.

7.1.1 Fuzzy Partitions

Fixing a fuzzy partition is the first step for F-transform based computations.

Let $I = [x_0, x_{m+1}]$ be a closed interval and $\{x_1, \ldots, x_m\}$, with $m \geq 3$, be points of I, called nodes, such that $x_0 < x_1 < \ldots < x_{m+1}$. A fuzzy partition of I is defined as a sequence $\{A_1, A_2, \ldots, A_m\}$ of fuzzy sets $A_i : I \to [0, 1]$, with $i = 1, \ldots, m$, identified by their membership functions on I such that

- $A_i(x) = 0$ if $x \notin (x_{i-1}, x_{i+1})$ (locality);
- A_i is continuous (continuity);
- $A_i(x) > 0$ if $x \in (x_{i-1}, x_{i+1})$ (positivity), with a unique maximum at x_i, where $A_i(x_i) = 1$;
- $\sum_{i=1}^{m} A_i(x) = 1, \quad \forall x \epsilon I$ (Ruspini's condition).

The functions $A_1(x), A_2(x), \ldots, A_m(x)$ are also called basic functions.

This is a more formal definition of fuzzy partition introduced in Chap. 2.

The basic functions form a h–*uniform* fuzzy partition, when the nodes are equidistant, with norm of the partition $h = (x_0 - x_{m+1})/(m + 1)$.

© The Author(s), under exclusive license to Springer Nature Switzerland AG 2022
S. Tomasiello et al., *Contemporary Fuzzy Logic*, Big and Integrated
Artificial Intelligence 1, https://doi.org/10.1007/978-3-030-98974-3_7

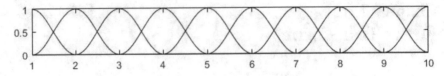

Fig. 7.1 Uniform fuzzy partitions by raised cosine shaped basic functions

For a h-uniform fuzzy partition, there exists an even function $A : [-1, 1] \rightarrow [0, 1]$ such that for all $k = 1, \ldots, m$, it is

$$A_k(x) = A\left(\frac{x - x_k}{h}\right), \ x \in [x_{k-1}, x_{k+1}]. \tag{7.1}$$

The function A_0 is called *generating function* of the uniform fuzzy partition. Two typical choices for A are:

- the identity function, giving the well-known triangular basic functions,
- $A = \frac{1}{2}(\cos(.) + 1)$, giving the raised cosine basic functions.

The raised cosine shaped function is defined as follows:

$$A_j(x) = \begin{cases} \frac{1}{2}\left(\cos\frac{\pi(x-x_j)}{(x_j-x_{j-1})} + 1\right), \ x \in [x_{j-1}, x_j] \\ \frac{1}{2}\left(\cos\frac{\pi(x-x_j)}{(x_{j+1}-x_j)} + 1\right), \ x \in [x_j, x_{j+1}] \\ \qquad\qquad\qquad 0, \quad otherwise. \end{cases} \tag{7.2}$$

Figure 7.1 shows a uniform fuzzy partition by raised cosine shaped basic functions.

7.1.2 F-Transform

Definition 7.1 The F-transform of a function $f(x)$, continuous on I, with respect to $\{A_1, A_2, \ldots, A_m\}$ is the m–tuple $[F_1, F_2, \ldots, F_m]$, whose ith element is

$$F_i = \frac{\int_{x_1}^{x_m} f(x)A_i(x)dx}{\int_{x_1}^{x_m} A_i(x)dx}. \tag{7.3}$$

Definition 7.2 The inverse F–transform is defined as the function

$$f^F(x) = \sum_{i}^{m} F_i A_i(x), \qquad x \in I. \tag{7.4}$$

In most cases, the function f is known only at a given set of points $x_j \in I$, $j = 1, \ldots, p$. In such a case, the discrete F-transform can be used.

Definition 7.3 The discrete F–transform is given by

$$F_i = \frac{\sum_{j=1}^{p} f(x_j) A_i(x_j)}{\sum_{j=1}^{p} A_i(x_j)}, \qquad i = 1, \ldots, m, \tag{7.5}$$

where the points $x_j \in I$ are such that for each $i \in \{1, \ldots, m\}$, there exists $\eta \in \{1, \ldots, p\}$, with $x_\eta \in supp(A_i)$, i.e., the set of points $\{x_1, x_2, \ldots, x_p\}$ is sufficiently dense with respect to the fuzzy partition.

By using the matrix notation, the discrete F-transform of an M-dimensional vector $\mathbf{v}$ is the m-dimensional vector ($m < M$)

$$\mathbf{F} = \mathbf{v}^T \mathbf{A} \overline{\mathbf{S}}^0, \tag{7.6}$$

where $\mathbf{A}$ is the $M \times m$ matrix

$$\mathbf{A} = \begin{pmatrix} A_1(x_1) & \ldots & A_m(x_1) \\ A_1(x_2) & \ldots & A_m(x_2) \\ \vdots & \vdots & \vdots \\ A_1(x_M) & \ldots & A_m(x_M) \end{pmatrix}, \tag{7.7}$$

$\overline{\mathbf{S}}^0$ is the inverse of the diagonal matrix with order m, whose non-null entries are given by $S_{jj} = \sum_{i=1}^{M} A_j(x_i)$. The discrete inverse F–transform provides the approximate solution, that is

$$\mathbf{v}^F = \mathbf{F} \mathbf{A}^T. \tag{7.8}$$

7.1.3 F-Transform in Two Variables

Let $f(x, y)$ be a continuous function on $[a, b] \times [c, d]$. The F-transform of $f(x, y)$, with respect to the fuzzy partitions $\{A_1, \ldots, A_n\}$ and $\{B_1, \ldots, B_m\}$ of the intervals $[a, b]$ and $[c, d]$ respectively, results in the $n \times m$ matrix, whose entries are computed as follows, for any $x \in [a, b]$ and $y \in [c, d]$:

$$F_{kl} = \frac{\int_c^d \int_a^b f(x, y) A_k(x) B_l(y) dx dy}{\int_c^d \int_a^b A_k(x) B_l(y) dx dy}, \tag{7.9}$$

with $k = 1, \ldots, n, l = 1, \ldots, m$. The inverse F-transform of $f(x, y)$, with respect to the fuzzy partitions $\{A_1, \ldots, A_n\}$ and $\{B_1, \ldots, B_m\}$ is the following function on $[a, b] \times [c, d]$

$$f^F = \sum_{k=1}^{n} \sum_{j=1}^{m} F_{kl} A_k(x) B_l(y). \tag{7.10}$$

The discrete F-transform of $f(x, y)$ with respect to the above-mentioned partitions is given by

$$F_{kl} = \frac{\sum_{j=1}^{M} \sum_{i=1}^{N} f(x_i, y_j) A_k(x_i) B_l(y_j)}{\sum_{j=1}^{M} \sum_{i=1}^{N} A_k(x_i) B_l(y_j)}, \qquad (7.11)$$

with $k = 1, \ldots, n, l = 1, \ldots, m$.

7.2   Types of Fuzzy Transform

The inverse F-transform can approximate the original continuous function (or data in the discrete case) with an arbitrary precision. Improved results are even possible through higher degree F-transform, denoted as F^s-transform, with $s \geq 0$. The F^s-transform components are polynomials of degree s, allowing for a better approximation. The quality of approximation improves as the degree of the polynomial increases. Such polynomials are orthogonal projections of an original function onto a linear subspace of functions with an inner product. When the degree of the polynomial projections is zero, the original F-transform, also denoted as F^0-transform, is retrieved. Here we recall only the F^1–transform (Hurtik and Tomasiello 2019).

7.2.1 F^1–Transform

Let $\mathbf{v}$ be an M–dimensional vector, whose entries can be regarded as given functional values $g(z_i)$, with $i = 1, \ldots, M$. The discrete F^1–transform of the vector $\mathbf{v}$ is the m–dimensional vector, whose elements are:

$$F_k^1(x_i) = c_k^0 + c_k^1(x_i - x_k), \qquad (7.12)$$

where

$$c_k^0 = \frac{\sum_{i=1}^{M} v_i A_k(x_i)}{\sum_{i=1}^{M} A_k(x_i)}, \quad c_k^1 = \frac{\sum_{i=1}^{M} v_i A_k(x_i)(x_i - x_k)}{\sum_{i=1}^{M} A_k(x_i)(x_i - x_k)^2}. \qquad (7.13)$$

The compact form of the inverse discrete F^1–transform is:

$$\mathbf{v}^{F,1} = \mathbf{c}^0 \mathbf{A}^T + \mathbf{c}^1 \overline{\mathbf{A}}^T, \qquad (7.14)$$

with the m-dimensional vectors

$$\mathbf{c}^0 = \mathbf{F}^0 = \mathbf{v}^T \mathbf{A} \overline{\mathbf{S}}^0, \quad \mathbf{c}^1 = \mathbf{v}^T \mathbf{A} \overline{\mathbf{S}}^1, \qquad (7.15)$$

where $\overline{\mathbf{A}}$ is the $M \times m$ matrix with entries $A_{ik} = A_k(x_i)(x_i - x_k)$ and $\overline{\mathbf{S}}^1$ is the inverse of the diagonal matrix whose non-null entries are $S_{kk}^1 = \sum_{i=1}^{M} A_k(x_i)(x_i - x_k)^2$.

In general, one can write

$$\mathbf{v}^{F,r} = \mathbf{c}^0 \mathbf{A}^T + \delta_{1r} \mathbf{c}^1 \overline{\mathbf{A}}^T, \tag{7.16}$$

where δ_{1r} is the Kronecker delta and $r \in \{0, 1\}$, so that when $r = 0$ the F^0-transform is retrieved.

The discrete F^1 transform of the $N \times M$ data matrix $\mathbf{D}$ is the $n \times m$ matrix, whose entries are computed as follows:

$$F_{kl}^1(i, j) = c_{kl}^{00} + c_{kl}^{10}(i - k) + c_{kl}^{01}(j - l), \tag{7.17}$$

where

$$c_{kl}^{00} = \frac{\sum_{j=1}^{M} \sum_{i=1}^{N} D_{ij} A_k(i) B_l(j)}{\sum_{j=1}^{M} \sum_{i=1}^{N} A_k(i) B_l(j)}, \tag{7.18}$$

$$c_{kl}^{10} = \frac{\sum_{j=1}^{M} \sum_{i=1}^{N} D_{ij} A_k(i)(i - k) B_l(j)}{\sum_{j=1}^{M} \sum_{i=1}^{N} A_k(i)(i - k)^2 B_l(j)}, \tag{7.19}$$

$$c_{kl}^{01} = \frac{\sum_{j=1}^{M} \sum_{i=1}^{N} D_{ij} A_k(i) B_l(j)(j - l)}{\sum_{j=1}^{M} \sum_{i=1}^{N} A_k(i) B_l(j)(j - l)^2}. \tag{7.20}$$

Thus, the inverse discrete F^1-transform for the bivariate case is

$$D_{ij}^{F,1} = \sum_{k=1}^{m} \sum_{l=1}^{n} F_{kl}^1 A_k(i) B_l(j). \tag{7.21}$$

In matrix form, (7.21) can be written as follows:

$$\mathbf{D}^{F,1} = \mathbf{A}\mathbf{C}^{00}\mathbf{B}^T + \overline{\mathbf{A}}\mathbf{C}^{10}\mathbf{B}^T + \mathbf{A}\mathbf{C}^{01}\overline{\mathbf{B}}^T, \tag{7.22}$$

where $\mathbf{C}^{00} = \mathbf{F}^0$, $\overline{\mathbf{A}}$ is an $N \times n$ matrix with entries $\overline{A}_{ik} = A_k(i)(i - k)$, $\overline{\mathbf{B}}$ is an $M \times m$ matrix with entries $\overline{B}_{jl} = B_l(j)(j - l)$ and

$$\begin{aligned} \mathbf{C}^{10} &= \mathbf{P}^{10} o \overline{\mathbf{Q}}^{10}, \\ \mathbf{C}^{01} &= \mathbf{P}^{01} o \overline{\mathbf{Q}}^{01}, \end{aligned} \tag{7.23}$$

with

$$\begin{aligned} \mathbf{P}^{10} &= \overline{\mathbf{A}}^T \mathbf{D}\mathbf{B}, \\ \mathbf{P}^{01} &= \mathbf{A}^T \mathbf{D}\overline{\mathbf{B}}, \end{aligned} \tag{7.24}$$

where $\overline{\mathbf{Q}}^{10}$, and $\overline{\mathbf{Q}}^{01}$ are two $n \times m$ matrices with entries $\overline{Q}_{ij}^{10} = (\sum_{j=1}^{M} \sum_{i=1}^{N} A_k(i)$
$(i-k)^2 B_l(j))^{-1}$ and $\overline{Q}_{ij}^{01} = (\sum_{j=1}^{M} \sum_{i=1}^{N} A_k(i) B_l(j)(j-l)^2)^{-1}$ respectively.
Finally, (7.22) can be written in a general form as follows

$$\mathbf{D}^{F,r} = \mathbf{A}\mathbf{C}^{00}\mathbf{B}^T + \delta_{1r} \left(\overline{\mathbf{A}}\mathbf{C}^{10}\mathbf{B}^T + \mathbf{A}\mathbf{C}^{01}\overline{\mathbf{B}}^T \right), \tag{7.25}$$

where δ_{1r} is the Kronecker delta and $r \in \{0, 1\}$.

7.2.2   Least–Squares Based F^0-Transform

In the least–squares (LS) approach, the elements of the discrete F–transform of f
with respect to $\{A_1, \ldots, A_m\}$ are obtained by minimizing the reconstruction error
(Patané 2011), that is they are handled as unknowns $\overline{\lambda}_i$, collected into the vector $\overline{\Lambda}$,
and appearing in the error vector $\mathbf{E}$:

$$\mathbf{E} = \mathbf{v} - \mathbf{A}\overline{\Lambda}. \tag{7.26}$$

By minimizing $\mathbf{E}$ with respect to the unknowns $\overline{\lambda}_i$, then

$$\overline{\Lambda} = \mathbf{K}^{-1}\mathbf{A}^T\mathbf{v}, \tag{7.27}$$

where

$$\mathbf{K} = \mathbf{A}^T\mathbf{A}. \tag{7.28}$$

The reconstructed vector is given by:

$$\mathbf{v}^{LS} = \mathbf{A}\overline{\Lambda}. \tag{7.29}$$

Since $\mathbf{K}$ is a Gram matrix, it is positive definite. If the membership functions are
generated by means of Gaussian or compactly-supported kernels, the matrix $\mathbf{K}$ turns
out to be sparse. There are good algorithms for the reduction of sparse matrices.
Anyhow, it is desirable working with band matrices. The computational cost of the
inversion of a symmetric banded matrix of order n can be reduced to $O(n)$. By means
of triangular or raised cosine shaped basic functions, $\mathbf{K}$ turns out to be a band matrix.
For the bivariate case, the error to be minimized is

$$\mathbf{E} = \mathbf{D} - \mathbf{A}\Lambda\mathbf{B}^T. \tag{7.30}$$

By the minimization of such functional with respect to λ_{ij}, one gets

$$\Lambda = \mathbf{K}^{-1}\mathbf{G}\mathbf{H}^{-1}, \tag{7.31}$$

where

$$\mathbf{G} = \mathbf{A}^T \mathbf{D} \mathbf{B},$$
$$\mathbf{K} = \mathbf{A}^T \mathbf{A}, \qquad (7.32)$$
$$\mathbf{H} = \mathbf{B}^T \mathbf{B}.$$

Data are reconstructed by

$$\mathbf{D}^{LS} = \mathbf{A} \Lambda \mathbf{B}^T. \qquad (7.33)$$

7.3 Application to Data Compression

Data compression can be regarded as a mean for representing information in compact form.

Typical application context is Wireless Sensors Networks (WSNs). In WSNs, data transmission turns out to be a critical operation in terms of energy waste and data compression is employed to address this issue.

It is usual to distinguish between two classes of compression techniques:

- lossless, where the reconstructed data is identical to the original data (prior to the compression);
- lossy, where the reconstructed data is affected by a certain error.

The F-transform based compression is a lossy technique.

Let $\mathbf{D}$ denote an $N \times M$ matrix. Such matrix with its entries can be identified by a function f,

$$f : J_{N,M} \rightarrow I \subset \mathbb{R}, \qquad (7.34)$$

where $J_{N,M} = \{1, 2, \ldots, N\} \times \{1, 2, \ldots, M\}$. is a finite rectangular domain.

One has to refer to the real intervals $[1, N]$ and $[1, M]$ in order to get the related fuzzy partitions. It is also worth mentioning that in the case of 8 bit gray-scale image, I is the finite interval $\{0, 1, \ldots, 255\}$.

The discrete F–transform of $\mathbf{D}$, with respect to the fuzzy partitions $\{A_1, \ldots, A_n\}$ and $\{B_1, \ldots, B_m\}$ of the intervals $[1, N]$ and $[1, M]$ respectively, with $n < N$ and $m < M$, can be written as

$$F_{kl} = \frac{P_{kl}}{Q_{kl}} \qquad k = 1 \ldots, n \quad l = 1, \ldots, m, \qquad (7.35)$$

where P_{kl} and Q_{kl} are the entries of the matrices $\mathbf{P}$ and $\mathbf{Q}$ respectively. These matrices are computed as follows

$$\mathbf{P} = \mathbf{A}^T \mathbf{D} \mathbf{B}, \quad \mathbf{Q} = \mathbf{A}^T \bar{\mathbf{I}}_{NM} \mathbf{B}, \qquad (7.36)$$

where $\mathbf{A}$ and $\mathbf{B}$ are the matrices with entries $A_k(i)$ and $B_l(j)$ respectively, $\bar{\mathbf{I}}_{NM}$ is the $N \times M$ matrix with all unit entries.

Fig. 7.2 Data compression
and decompression by
F-transform

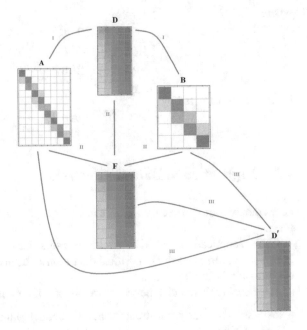

The decompression of the matrix **D** is obtained by the discrete inverse F–transform
as follows

$$\mathbf{D}^F = \mathbf{AFB}^T. \tag{7.37}$$

By introducing the matrix $\overline{\mathbf{Q}}$, whose entries are $\overline{Q}_{ij} = 1/Q_{ij}$, then one can get
the compact form of (7.35)

$$\mathbf{F} = \mathbf{P}o\overline{\mathbf{Q}}, \tag{7.38}$$

where o denotes the Hadamard product.
   Figure 7.2 visualizes the whole process:

- stage I, the matrices **A** and **B** are generated from the given data matrix **D**;
- stage II, the matrices **A**, **B** and **D** are used to obtain the matrix **F** of the fuzzy
  transform elements;
- stage III, the matrices **A**, **B** and **F** are employed to get the reconstructed data matrix
  $\mathbf{D}^F$.

7.3.1 F^0-Transform Code

Let us consider the compression in one direction, e.g. to reduce the number of
instances N of a data set, by fixing $n < N$. This can be useful to preprocess high-
dimensional data sets (e.g. when there is a certain correlation), before they are presented to a neural network to reduce the training time. The technique is embodied

in the function `[rmse,RX,F]=ft(X,n)`. This function returns the RMSE, the compressed matrix F and the reconstructed data RX, given a data matrix X, by means of raised cosine type uniform fuzzy partition with dimension n (compression enabled only in one direction).

The nodes of the fuzzy uniform partition are first obtained

```
for i=1:n
D(i)=(i - 1)/(n - 1)*(ss(1) - 1) + 1;
end
```

where `ss=size(X);` has been initially set.

The norm of the uniform partition is easily computed

```
h=(D(n)-D(1))/(n-1);
```

After initializing the matrix A of the membership grades, the fuzzy partition is obtained

```
for k=1:ss(1)
if D(1)<=k & k<=D(2)
A(k,1)=0.5*(cos(%pi*(k-D(1))/h)+1);
else
if D(n-1)<=k & k<=D(n)
A(k,n)=0.5*(cos(%pi*(k-D(n))/h)+1);
end
end
for j=2:n-1
if D(j-1)<=k & k<=D(j+1)
A(k,j)=0.5*(cos(%pi*(k-D(j))/h)+1);
end
end
end
```

Finally, it is quite easy to set the direct and inverse F-transform formulas

```
M1=A'*X*B;
M2=A'*I*B;
F=M1./M2;
RX=A*F*B';
```

It is worth mentioning that here B is an identity matrix ($m = M$), i.e.

```
B=eye(ss(2));
```

A complete listing appears in the Appendix.

Example 7.1 The data set here considered is Communities and crime (CC), from the UCI machine learning repository (http://archive.ics.uci.edu/ml/datasets/communities+and+crime). The original data set presents 1994 instances, 122 predictive attributes. In the performed numerical experiment, only the first 96 attributes have been considered, discarding the remaining ones because of the missing values.

In what follows, comm.txt is the file containing the data matrix.

Fig. 7.3 Scilab console: numerical experiments with the F-transform

```
--> S=fscanfMat('comm.txt');

--> for i=400:400:1600 rmse=ft(S,i) end
rmse  =

  0.1098869

rmse  =

  0.0968816

rmse  =

  0.0817768

rmse  =

  0.0654575
```

By typing in the Scilab console:

```
S=fscanfMat('comm.txt');
for i=400:400:1600 rmse=ft(S,i) end
```

to try different compression ratios, one gets the results (RMSE) depicted in Fig. 7.3. It is worth noticing that the error (RMSE) in the reconstructed data decreases for lower compression ratios, i.e. by increasing n.

7.4   Application to Machine Learning Algorithms

The fuzzy transform, meant as F^0-transform, can be used to lower the computational burden of computing schemes, such as ANNs. By using the F-transform as a reduction technique, e.g. to reduce the number of columns or rows of the given data set, intuitively, the training time gets shorter.

In order to support the discussion, it is useful to recall that in a learning system there are training time and query time. Training time is the time employed by the system to learn from a training data set to make inferences. The query time is when some specific data (not belonging to the training data set) are presented to the system and a result is returned.

In what follows, we refer to a computing scheme based on Lazy Learning (LL). This is an instance-based learning method. This means that once a query is received, the prediction is carried out by means of a local regression model with respect to the neighbouring examples of the query, properly selected on the basis of a distance measure. Unlike ANNs, LL does not create an abstraction from the given instances (Bontempi 1999). The data is stored, and at query time, an answer is derived on the basis of the query's nearest neighbors. Hence, applying a reduction technique, such as the F-transform, is critical to the storage burden.

7.4.1 Local Weighted Regression

We briefly recall the algorithm proposed by Birattari et al. (1999).

Let $\mathbf{X}_i^T$ and $\mathbf{X}_q$ be the row vectors of the data matrix $\mathbf{X}$ and the query vector respectively. Then the algorithm can be summarized as follows

- compute the Euclidean distances

$$d_{iq} = \|X_i - X_q\|, \qquad i = 1, \dots, N \tag{7.39}$$

- define the weights

$$w_i = \sqrt{K(d_{iq}, \overline{h})}, \qquad i = 1, \dots, N \tag{7.40}$$

with

$$K(d_{iq}, \overline{h}) = \begin{cases} 1 & if \quad d_{iq} \leq \overline{h} \\ 0 & otherwise \end{cases} \tag{7.41}$$

where $\overline{h}$ is the bandwidth, to be properly fixed;

- generate the matrix $\mathbf{M}$ and the vector $\overline{\mathbf{v}}$ as follows

$$\mathbf{M} = \mathbf{WX}, \quad \overline{\mathbf{v}} = \mathbf{WY}, \tag{7.42}$$

where $\mathbf{W}$ is the diagonal matrix whose non-null entries are w_i and $\mathbf{Y}$ is the observed output vector;
- solve the linear regression problem

$$\mathbf{V}\beta = \mathbf{M}^T\overline{\mathbf{v}}, \tag{7.43}$$

where $\mathbf{V} = \mathbf{M}^T\mathbf{M}$ and β is the vector of unknowns;
- compute the desired prediction value with respect to the query vector X_q

$$\overline{y}_q = \mathbf{X}_q^T \beta = \mathbf{X}_q^T \mathbf{V}^{-1}\mathbf{M}^T\overline{\mathbf{v}}. \tag{7.44}$$

A locally weighted regression can be cast into the form of a linear smoother. A linear smoother is a model where the prediction is represented by a linear function of the observed output vector $\mathbf{Y}$, that is

$$\overline{\mathbf{Y}} = \mathbf{TY}, \tag{7.45}$$

where $\mathbf{T} = \mathbf{X}(\mathbf{X}^T\mathbf{W}^T\mathbf{WX})^{-1}\mathbf{X}^T\mathbf{W}^T\mathbf{W}$. The smoothness of the estimate depends on the smoothing parameter, that is the bandwidth $\overline{h}$. Usual methods for selecting the smoothing parameter are cross-validation and its variants and the Akaike Information Criterion (AIC). Cross-validation techniques are based on the idea that the model can be built from a part of the data, while the remainder of the data can be used to check the predictions.

A cross-validation technique to estimate the generalization error with respect to a finite set of data is the Leave-One-Out (LOO) (Birattari et al. 1999). The LOO error referred to the jth vector $\mathbf{X}_j$, such that $d_{jq} \leq \overline{h}$, is given by

$$e_j^{LOO} = (y_j - \mathbf{X}_j^T\beta)(1 - \mathbf{X}_j^T\mathbf{V}^{-1}\mathbf{X}_j)^{-1}, \tag{7.46}$$

or equivalently, in matrix form:

$$\mathbf{e}^{LOO} = \mathbf{D}^{-1}(\mathbf{Y} - \mathbf{X}\beta), \tag{7.47}$$

where $\mathbf{D}$ is the diagonal matrix whose entries are $D_{jj} = 1 - \mathbf{X}_j^T\mathbf{V}^{-1}\mathbf{X}_j$.

7.4.2 Lazy Learning with F-Transform

F-transform can be used to reduce both the storage burden due to the historical data and the running time of the learning algorithm (Loia et al. 2020). F-transform is applied to the $N \times M$ input data matrix $\mathbf{X}$ and the N-sized output vector $\mathbf{Y}$, in order to get the $n \times m$ compressed matrix $\mathbf{F}^X$ and the n-sized vector $\mathbf{F}^Y$. Hence, $\mathbf{F}^X$ and $\mathbf{F}^Y$ are used instead of $\mathbf{X}$ and $\mathbf{Y}$ respectively in (7.39), (7.42) in order to achieve a local regression model. The prediction value corresponding to the query vector $\mathbf{X}_q$ is obtained through the resulting (7.44). The LOO error is computed as in (7.46), but replacing the vectors $\mathbf{X}_j$ with the analogous vectors $\mathbf{F}_j^X$.

By using F-transform, the LOO error can be written as follows

$$\overline{\mathbf{e}}^{LOO} = \overline{\mathbf{M}}^{-1}(\mathbf{F}^Y - \mathbf{F}^X\beta) = \overline{\mathbf{M}}^{-1}\mathbf{S}^{-1}\mathbf{A}(\mathbf{Y} - \mathbf{X}\beta). \tag{7.48}$$

The approach consists of two stages:

- first the historical data are preprocessed by means of F-transform, before running the local regression algorithm;
- as soon as a query vector is presented, the prediction formula is invoked.

Example 7.2 Predicting chaotic time series
Let us consider the time series generated by the chaotic Mackey–Glass time-delay differential equation, examined in the Example 6.1. The aim is again to predict the value $x(t + 6)$ by considering the four input values $x(t - 18), x(t - 12),$

$x(t - 6), x(t)$. For comparative purposes, we use the same data set. Thus, the first 500 data samples were used as the training data set and the remaining 500 samples as the test data set. It is worth mentioning that this data set is different from the one used by Loia et al. (2020).

The code for LL has been imported in Scilab, converting the original one written by Birattari et al. (1999).

By means of LL–FT (with $N = 500$ and $n = 300$), it is possible to get the testing result $RMSE = 9.58E - 02$, which is similar to the ones reported in Example 6.1. Anyway, the training time by ANFIS-T turns out to be the shortest one: 0.06 s against 0.11 s by LL-FT.

Problems

1. Consider Example 7.1. Implement the triangular basic function in the code. Run the program and see how the results change.
2. Consider again Example 7.1. Make a few changes in the code to get a function for the compression in both directions and see how the results change.
3. Consider Example 7.2. Perform some numerical experiments by using ANNs, by preprocessing data by F^0-transform, as described in the same example.

References

Birattari M, Bontempi G, Bersini H (1999) Lazy learners at work: the lazy learning toolbox. In: Proceedings of 7th European congress on inteligent techniques and soft computing (EUFIT'99)

Bontempi G (1999) Local learning techniques for modelling, prediction and control. IRIDIA - Universite Libre de Bruxelles, Ph.D. thesis

Hurtik P, Tomasiello S (2019) A review on the application of fuzzy transform in data and image compression. Soft Comput 23(23):12641–12653

Loia V, Tomasiello S, Vaccaro A, Gao J (2020) Using local learning with fuzzy transform: application to short term forecasting problems. Fuzzy Optim Decis Making 19:13–32

Patané G (2011) Fuzzy transform and least-squares approximation: analogies, differences, and generalizations. Fuzzy Sets Syst 180(1):41–54

Perfilieva I (2006) Fuzzy transforms: theory and applications. Fuzzy Sets Syst 157:993–1023

Chapter 8
Introduction to Granular Computing

8.1 Fundamentals of Granular Computing

Granular Computing is about representing, constructing, and processing information granules. This paradigm has been conceived to be human centric. This is mainly possible thanks to the level of abstraction realized through the information granules, because humans perceive the world, reason, and communicate at a certain level of abstraction. Abstraction can be pictured with nonnumeric constructs, such as a collection of entities characterized by some notions of closeness, proximity, resemblance, or similarity. These collections are referred to as information granules. From a practical application standpoint, human centricity is a feature of intelligent systems and information granules function in the context of intelligent systems.

In the context of computational intelligence, the granular approaches overlap other techniques such as neural networks and evolutionary techniques (Fig. 8.1), this is because the main goal of the granular paradigm is to provide more transparent and efficient architectures and the evolutionary computation may be involved for the optimization of some parameters.

8.1.1 Information Granules

The concept of granulation and information granules dates back to a former work by Zadeh.

Granulation of an object $\overline{A}$ leads to a collection of granules of $\overline{A}$, with a granule being a clump of points (objects) drawn together by indistinguishability, similarity, proximity or functionality (Zadeh 1997).

An information granule defined in a space X can be seen as a mapping

$$A : X \rightarrow \mathcal{G}(x) \tag{8.1}$$

© The Author(s), under exclusive license to Springer Nature Switzerland AG 2022
S. Tomasiello et al., *Contemporary Fuzzy Logic*, Big and Integrated
Artificial Intelligence 1, https://doi.org/10.1007/978-3-030-98974-3_8

Fig. 8.1 In the context of
computational intelligence

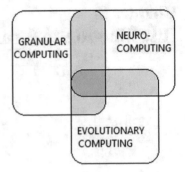

where A is an information granule, $\mathcal{G}$ denotes a formal framework of information granules.

Formal models of information granules can be built by means of

- set theory and interval analysis,
- fuzzy sets,
- rough sets,
- probabilistic sets.

The level of abstraction supported by the information granules is associated with the number of elements in the granule. The cardinality is used to determine the number of such elements. In general, the information granularity can be quantified through the integral (assuming that it exists)

$$Card(A) = \int_X \mathcal{F}(A(x))dx \qquad (8.2)$$

where A is an information granule and $\mathcal{F}$ is a certain monotonically increasing transformation of A.

By assuming $\mathcal{F}$ as the identity function and for finite universe of discourse, then (8.2) becomes

$$Card(A) = \sum_{i=1}^{n} A(x_i).$$

The two criteria guiding the formation of an information granule A are

- *coverage*, that is a quantification of how many data are "covered" by A; the higher the coverage, the more justified (by data) A is;
- *specificity*, in order to retain its semantics, the information granule should be as specific (detailed) as possible.

Formally, the coverage is expressed as the cardinality of the information granule. The specificity is related to the level of detail entailed by the granule, i.e. the support length of the granule. The higher the cardinality, the higher the abstraction of the granule and the lower its specificity.

8.1.2 Granularity and Fuzzy Sets

Fuzzy sets can be characterized by two scalar indexes describing their size (granularity) and uncertainty (viewed from the standpoint of hesitation in assigning elements to the fuzzy set). The first indicator is congruent with the cardinality of sets and it is represented by the energy measure of fuzziness; the second one is related to the notion of entropy as a measure of choice (selection).

The *energy measure of fuzziness* of the fuzzy subset A of X, denoted by $E(A)$, is defined as follows

$$E(A) = \int_X g(A(X))dx, \tag{8.3}$$

where $g : [0, 1] \to [0, 1]$ is an increasing function.

This index is about counting elements in the fuzzy set (this is clear in the discrete case). The simplest case is given by g as the identity function, leading to cardinality.

The *entropy measure of fuzziness* $H(.)$ can be regarded as a way of expressing hesitation as to the membership grades of the individual elements belonging to the fuzzy set.

This measure is similar to the energy measure of fuzziness:

$$H(A) = \int_X h(A(X))dx, \tag{8.4}$$

where $h : [0, 1] \to [0, 1]$ is a continuous function satisfying the property of symmetry $h(u) = h(1 - u)$ and monotonicity, that is $h(u)$ is monotonically increasing over $[0, 1/2]$ and monotonically decreasing over $[1/2, 1]$, with the boundary conditions $h(0) = h(1) = 0$ and $h(1/2) = 1$.

In order to quantify the hesitation of selecting a certain element from the fuzzy set to be treated as its representative, there is the *measure of specificity*, $Sp(.)$ that maps A into a single number $Sp(A)$ given as the integral

$$Sp(A) = \int_0^{\alpha_{max}} \frac{d\alpha}{Card(A_\alpha)}, \tag{8.5}$$

where α_{max} is the height of A.

It is clear that if A consists of a single element, then there is no hesitation, but if A has as many elements as needed to cover X, then there is the highest level of hesitation.

8.1.3 Construction of Information Granules Through Clustering

Clustering is a natural way to construct information granules in the presence of numeric data (Pedrycz 2016). There are different clustering algorithms. One of the most popular clustering algorithms, producing a collection of fuzzy sets, is fuzzy c-means (FCM). The method can be regarded as a generalization of the K-means clustering technique that instead can form set-based information granules. Here, we briefly recall the FCM steps.

Given a collection of p-dimensional data samples $\{\mathbf{x}_k\}$, $k = 1, 2, \ldots, N$, the task of determining its structure (a collection of "c" clusters) is expressed as the minimization of the following objective function (performance index) Q, written as a sum of the squared distances

$$Q = \sum_{i=1}^{c} \sum_{k=1}^{N} u_{ki}^{m} \|\mathbf{x}_k - \mathbf{v}_i\|^2, \tag{8.6}$$

where $\mathbf{v}_1, \mathbf{v}_2, \ldots, \mathbf{v}_c$ is a collection of c p-dimensional prototypes of the clusters and $\mathbf{U} = [u_{ki}]$ is a partition matrix, with u_{ki} the membership degree of data $\mathbf{x}_k$ in the ith cluster. The coefficient m (any real number greater than 1) expresses the impact of the membership grades on the individual clusters and produces a certain geometry of the information granules.

A partition matrix satisfies two properties:

$$0 < \sum_{k=1}^{N} u_{ki} < N, \quad i = 1, 2, \ldots, c, \tag{8.7}$$

$$\sum_{i=1}^{c} u_{ki} = 1, \quad k = 1, 2, \ldots, N. \tag{8.8}$$

Let $\mathcal{U}$ denote a family of matrices satisfying these two requirements. The minimization of Q with respect to $\mathbf{U} \in \mathcal{U}$ and the prototypes $\mathbf{v}_i$ has to take into account the constraints (8.7)–(8.8).

The use of Lagrange multipliers transforms the problem into its constraint-free version. The augmented objective function formulated for each data point, i.e. for $k = 1, 2, \ldots, N$, is

$$V_k = \sum_{i=1}^{c} u_{ki}^{m} \|\mathbf{x}_k - \mathbf{v}_i\|^2 + \lambda \left(\sum_{i=1}^{c} u_{ki} - 1 \right). \tag{8.9}$$

Hence, the successive entries of the partition matrix can be obtained as follows:

$$u_{ki} = \frac{1}{\sum_{j=1}^{c} \left(\frac{\|\mathbf{x}_k - \mathbf{v}_i\|}{\|\mathbf{x}_k - \mathbf{v}_j\|} \right)^{\frac{2}{m-1}}} \tag{8.10}$$

Finally, assuming the Euclidean distance between the data and the prototypes, that is $\|\mathbf{x}_k - \mathbf{v}_i\|^2 = \sum_{j=1}^{p} (x_{kj} - v_{ij})^2$, one obtains

$$v_{ij} = \frac{\sum_{k=1}^{N} u_{ki}^m x_{kj}}{\sum_{k=1}^{N} u_{ki}^m}, \tag{8.11}$$

with $j = 1, \ldots, p$. The procedure is iterated and (8.10), (8.11) updated. The procedure will stop at the sth iteration if $\|\mathbf{U}^{(s)} - \mathbf{U}^{(s-1)}\| < \epsilon$, where $0 < \epsilon < 1$ is the parameter of the termination criterion.

8.1.4 Another Type of Information Granule

In this section, a recent type of information granule is introduced, To this end, some notions have to be recalled.

Briefly, in stochastic theory, the available information is handled via probability density functions. In possibility theory, there is a possibility measure. The possibility function is a distribution of possibilities on the considered space Ω.

Definition 8.1 A possibility distribution on the set $\Omega \neq \emptyset$ is a function $\psi : \Omega \to [0, 1]$ satisfying

$$\sup_{\omega \in \Omega} \psi(\omega) = 1. \tag{8.12}$$

Formally, ψ is derived from a membership function of a fuzzy subset of Ω. Any normal fuzzy subset of Ω can be used to define a possibility distribution on Ω.

Definition 8.2 Let A_i be normal and convex fuzzy sets, for $i = 1, \ldots, m$. For any given vector $\mathbf{x}^T = (x_1, x_2, \ldots, x_n) \in \mathbb{R}^n$, the granule Γ^r is the sequence of the fuzzy sets A_i satisfying

$$\max_{x_j, j=1,\ldots,n, j \neq r} \psi_\Gamma(x_1, x_2, \ldots, x_n) = A_i(x_r), \tag{8.13}$$

where $\psi_\Gamma \in \mathbb{R}^n$ is a possibility distribution.

8.2   Granular Neural Networks

In the design of neural networks, the process of information granulation brings some advantages. In particular, it allows to deal with large amount of numerical details, usually involved in the form of fuzzy sets, with a more efficient learning.

The concept of granular neural network (GNN) was first formalized by Pedrycz and Vukovich in 2001. Over the last decade, several types of GNNs have been proposed (Song and Wang 2013). Depending on the extent of granulation, one can distinguish among four types of GNNs.

- *Training and testing with numeric (non-fuzzy) data.*
  These architectures converge to the standard ANNs if the data is non-fuzzy.
- *Training with numeric (non-fuzzy) data and testing on granular data.* A network may be trained using non-fuzzy data (high-granularity data) and then used in a low-granularity domain. The ANN has to be designed to accept granular input data during the testing.
- *Training with granular data and test on numeric (non-fuzzy) data.*
  In this case, the ANN is designed to process granular data and to process high-granularity data.
- *Training and testing with granular data.*
  This is the most general case.

Depending on whether they present modularity or not, GNNs can be categorized into two types.

- *GNNs with modular structure.* The data set is divided into sub-granules, each one with a different number of data. Independent neural networks serve as module and operate on separate inputs.
- *GNNs without modular structure.* The network functions as a single model.

The architecture of a single GNN may have granular connections or classical connections. Depending on this, GNNs can be categorized into two types.

- *GNNs with granular connections.* The weights of the network can be represented by fuzzy sets or other formalism congruent with the granulation. Thus, even in case of numeric input data, the output cannot be merely numeric.
- *GNNs without granular connections.* Only input data and output data are granulated. The values of the weights are numeric.

For instance, the granular neural network by Song and Pedrycz (2013) includes a single hidden layer with a certain number of neurons and one neuron in the output layer. The input data are non-fuzzy and the output data are intervals and the weights of the network are intervals as well. Hence, this model is not modular, with granular connections.

8.2.1 Granular Functional Networks and Granular Functional Link Artificial Neural Networks

The Granular Functional Network (GFN) and the Granular Functional Link Artificial Neural Network (GFLANN) are the granular version of the Functional Network (FN) and the Functional Link Artificial Neural Network (FLANN) with functional expansion, respectively. FN and FLANN with functional expansion have the same architecture, the difference is in the learning algorithm: LS-based for the FN and via backpropagation for the FLANN. Their typical architecture foresees

- a layer of input units (IL), handling the input signals;
- a layer of output units (OL), containing the output data;
- a layer of inner neurons (NL), processing a collection of input signals and providing a collection of signals by means of suitable functions.

A GFN, or similarly a GFLANN, has an intermediate layer, namely a granular layer (GL) between the layers IL and NL. It consists of a collection of fuzzy sets, meant as a fuzzy partition of the generic domain of the input variable.

As a kind of neural network, the FN can be regarded as a nonlinear dynamical system, modelled by

$$\dot{x}_j(t) = \overline{F}_j(x_j(t)), \quad j = 1, \ldots, n, \tag{8.14}$$

where $x_1, \ldots, x_n$ denote the state variables of the dynamical system and $\overline{F}_j(.)$ a nonlinear function of its argument.

Let $A_l(.)$ be m basic functions giving a fuzzy partition of the reference domain. The discretized version of (8.14), by using the approximation given by the basic functions is

$$x_j(s+1) = \sum_{k=1}^{n} \sum_{l=1}^{m} (-1)^k A_l(x_k(s)) w_{kl}^{(j)}, \quad j = 1, \ldots, n \tag{8.15}$$

with the weights $w_{kl}^{(j)} \in [0, 1]$, and s the discrete time.

A simple scheme of GFN, with only one output, is depicted in Fig. 8.2.

Both the GFN and the GFLANN are based on the information granule (8.13). For each granule, the approximation is:

$$\phi^l(A_l(x_1(s), x_2(s), \ldots, x_n(s))) = \bigvee_{k=1}^{n} (A_l(x_k(s)) * w_{kl}), \tag{8.16}$$

for $l = 1, \ldots, m$, where $*$ is a t-norm, $\bigvee$ is the maximum operator, $A_l(x_k(s))$ is the membership degree of the numerical data coming from the kth internal domain and $w_{kl} \in [0, 1]$ are the weights.

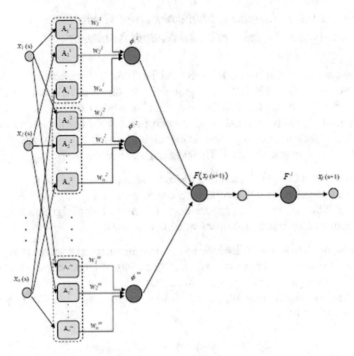

Fig. 8.2 A simple Granular Functional Network

In the general case, with more than one output, each output $y_j = x_j(s+1)$ is processed through some invertible functions F_j, that is $F_j(y_j)$, with $j = 1, \ldots, p$.

From the proposed scheme it is possible to extract Takagi-Sugeno-Kang like rules such as:

$$R_k^i : \text{IF} x_1 \text{is} A_i \text{AND} x_2 \text{is} A_i \text{AND} \ldots x_n \text{is} A_i \text{THEN} y_k = F_k^{-1}(\phi^i(x_1, \ldots, x_n)),$$

where $x_1, \ldots, x_n$ are non-fuzzy data, as mentioned before.

Regarding the learning algorithm, there are two types of weights: granular weights, denoted as w_{ij}, which are randomly chosen in $[0, 1]$, output weights, denoted as $\overline{w}_{lm}$, which are unknown. In the GFN scheme, they are learned through an LS approach as explained in the following.

Let N_p be the number of samples $\mathbf{x}^{(i)}$ in the training data. Let $\mathbf{A}$ be the $N_p \times mn$ matrix of the membership degrees of the input values in the fuzzy sets A_i:

$$\mathbf{A} = \begin{pmatrix} \mathbf{A}_1(\mathbf{x}^{(1)})^T & \ldots & \mathbf{A}_m(\mathbf{x}^{(1)})^T \\ \mathbf{A}_1(\mathbf{x}^{(2)})^T & \ldots & \mathbf{A}_m(\mathbf{x}^{(2)})^T \\ \vdots & \vdots & \vdots \\ \mathbf{A}_1(\mathbf{x}^{(N_p)})^T & \ldots & \mathbf{A}_m(\mathbf{x}^{(N_p)})^T \end{pmatrix}, \tag{8.17}$$

with the row vectors $\mathbf{A}_i(\mathbf{x}^{(j)})^T = (A_i(x_1^{(j)}), \ldots, A_i(x_n^{(j)}))$, where $x_k^{(j)}$ denotes the kth input value in the jth pattern.

Let $\mathbf{W}$ be the $mn \times m$ matrix of granular weights:

$$\mathbf{W} = \begin{pmatrix} \mathbf{w}_1 & \mathbf{0} & \mathbf{0} & \ldots & \mathbf{0} \\ \mathbf{0} & \mathbf{w}_2 & \mathbf{0} & \ldots & \mathbf{0} \\ \vdots & \vdots & \vdots & \vdots & \vdots \\ \mathbf{0} & \mathbf{0} & \ldots & \mathbf{0} & \mathbf{w}_m \end{pmatrix}, \tag{8.18}$$

where $\mathbf{0}$ is an n-sized vector with null elements and $\mathbf{w}_i^T = (w_{i1}, \ldots, w_{in})$. Finally, let $\mathbf{V}$ denote the $N_p \times p$ matrix of target values. Then it is

$$\mathbf{V} = \mathbf{P}\overline{\mathbf{W}}, \tag{8.19}$$

where $\mathbf{P} = \mathbf{A}\mathbf{W}$.

The unknown weights matrix $\overline{\mathbf{W}}$ can be computed as follows

$$\overline{\mathbf{W}} = \mathbf{M}^{-1}\mathbf{P}^T\mathbf{V}, \tag{8.20}$$

where $\mathbf{M} = \mathbf{P}^T\mathbf{P}$.

In a GFLANN, the following delta learning rule is used to adjust the weights at each iteration s:

$$\overline{w}_{ji}(s+1) = \overline{w}_{ji}(s) + \Delta\overline{w}_{ji}(s), \quad \Delta\overline{w}_{ji}(s) = -\eta\frac{\partial E}{\partial\overline{w}_{ji}}. \tag{8.21}$$

Here, $\eta \in (0, 1)$ is the learning rate to be fixed, and $E(s) = \frac{1}{2}\mathbf{e}^T(s)\mathbf{e}(s)$ is the error functional based on the distances between the computed values y_j and the target values $\overline{y}_j$, $e_j(s) = y_j(s) - \overline{y}_j(s)$.

$$\frac{\partial E}{\partial\overline{w}_{ji}}(s) = \frac{\partial E}{\partial e_j}\frac{\partial e_j}{\partial y_j}\frac{\partial y_j}{\partial\overline{w}_{ji}} = -e_j(s)\phi_i, \tag{8.22}$$

where $\phi_i = \sum_{k=1}^n w_{ki}A_i(x_k)$. Whence it follows that

$$\overline{w}_{ji}(s+1) = \overline{w}_{ji}(s) + \eta(y_j(s) - \overline{y}_j)\phi_i. \tag{8.23}$$

The convergence and the accuracy have been formally discussed by Colace et al. (2020).

8.2.2   A Granular Recurrent Neural Network

In this section the Granular Recurrent Neural Network (GRNN) introduced by
Tomasiello et al. (2021) is presented. This scheme is deduced from a generic system
of partial differential equations with the aim to capture the spatio-temporal variability
of some data sets and problems.

By means of several steps through differential quadrature and finite difference,
the discrete system as the following is obtained

$$U(s + 1) = U(s) + \rho V(s), \tag{8.24}$$
$$V(s + 1) = WV(s) + W^{BACK}U(s) + W^{IN}F(s), \tag{8.25}$$

where ρ is the spacing chosen for the time discretization, s the discrete time and
$F(s)$ represents the input vector. As a result of the adopted modelling, the following
matrices

$$W = I_{NP} - \rho W_0 \otimes I_N, \qquad W^{IN} = \rho W_{I0} \otimes I_N, \tag{8.26}$$

can be built by randomly choosing the matrices W_0 and W_{I0} in the range $(0, 1)$,
while $0 < \rho \ll 1$, and performing the Kronecker product with the identity matrix of
dimension N. The weights W^{BACK} are also randomly chosen in the range $(0, 1)$.

It is assumed that there exist $\mu_{ij} \in \mathbb{R}$, collected into the $NP \times mNP$ matrix M,
such that the equation to update U can be written as:

$$U(s + 1) = MB(V(s)), \tag{8.27}$$

where

$$B(V(s))^T = (A_1(v_{11}(s)), \ldots, A_m(v_{PN}(s))),$$

with $A_1, \ldots, A_m$ as basic functions forming a fuzzy partition of the interval $\overline{I} =
[v_{min}, v_{max}]$, with $v_{min} = \min_i V_i$, $v_{max} = \max_i V_i$. Here V_i, with $i = 1, \ldots, NP$, are
meant as internal states. The matrices W and W^{IN} are sparse matrices, in particular
when $P = L = 1$, they both are diagonal matrices. In this scheme, the number of
output and internal nodes is the same, but the internal states are processed by a
variable number of granules. By means of successive substitutions, one gets

$$V(s) = W^s V(0) + \sum_{i=0}^{s-1} W^i W^{BACK} U(s - i - 1) + \tag{8.28}$$

$$+ \sum_{i=0}^{s-1} W^i W^{IN} F(s - i - 1).$$

The matrices W and W^{IN} are kind of sparse matrices, consisting of P diagonal block matrices, whose dimension is N. When $P = L = 1$, they both are diagonal matrices. In general, they can also be regarded as kind of band matrices, whose bandwidth is $N(P - 1) + 1$. Working with band matrices is beneficial from a computational perspective. A band matrix can be likened in complexity to a rectangular matrix whose row dimension is equal to the bandwidth of the band matrix. Thus performing operations such as the multiplications in (8.28) leads to significant time savings, since creating the internal layer is a very time demanding task.

The granular approximation (8.27) is obtained by means of a granular layer between the internal and the output layers, as depicted in Fig. 8.3 (where $P = 1$ for convenience).

From a fuzzy mathematics perspective, for each information granule, based on (8.13), we have the following approximation (again with $P = 1$ for a better readability):

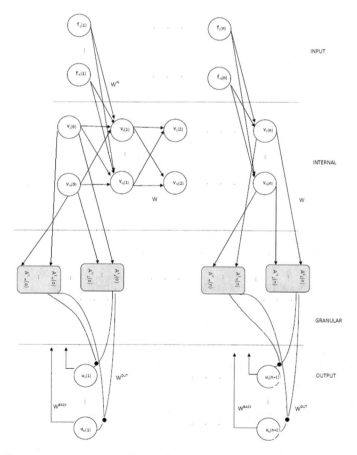

Fig. 8.3 The granular recurrent neural network

$$\phi^l(A_l(v_1(i), v_2(i), \ldots, v_N(i))) = \bigvee_{l=1}^{m} A_l(v_k(i)) * \omega_l, \tag{8.29}$$

where $*$ is a t-norm, $\bigvee$ is the maximum operator, $A_l(x_k(i))$ is the membership degree of the data from the kth internal domain and ω_l are the weights.

Hence, it is possible to extract from this scheme Takagi-Sugeno-Kang like rules such as:

$$R_j^s : \text{IF } v_1(i) \text{ is } A_s \text{ AND } v_2(i) \text{ is } A_s \text{ AND} \ldots v_N(i) \text{ is } A_s \text{ THEN } u_j(i+1) =$$
$$\overline{F}(\rho^s(A_s(v_1(i), \ldots, v_N(i)))),$$

where $\overline{F}$ may be any suitable function, here as the identity function.

Let us recall (8.27) now. By collecting the target values into the $n \times NP$ matrix T, and the vectors $B(V(i))^T$ into the $n \times mNP$ matrix $\overline{B}$, being n the maximum discrete time value considered, then we have

$$T = \overline{B} W^{OUT}, \tag{8.30}$$

where $\overline{B}^T = (B(V(0))^T, \ldots, B(V(n-1))^T)$, $T^T = (U(1)^T, \ldots, U(n)^T)$, and W^{OUT} the $mNP \times NP$ matrix to be determined.

In this case, by means of least–squares minimization, one gets

$$W^{OUT} = (\overline{B}^T \overline{B})^{-1} \overline{B}^T T. \tag{8.31}$$

In a more general way, one can rewrite (8.31) by using a regularization parameter λ (standard Tikhonov regularization), which helps in case of badly conditioned matrices:

$$W^{OUT} = (\overline{B}^T \overline{B} + \lambda \mathbf{I})^{-1} \overline{B}^T T. \tag{8.32}$$

Regarding the accuracy of the GRNN scheme, it is possible to prove that under some hypotheses the computed output equals the target as $m \to \infty$ (Tomasiello et al. 2021).

For the stability analysis one has to consider the following map

$$V(s+1) = G(V(s)), \tag{8.33}$$

with $G(V(s)) = WV(s) + W^{BACK} MB(V(s-1)) + W^{IN} F(s)$.

It is useful to recall that an equilibrium (or fixed) point $\overline{V}$ is such that $G(\overline{V}) = \overline{V}$. An equilibrium point is said to be asymptotically stable if all the eigenvalues of the Jacobian matrix $JG(V(s))$ have moduli less than one.

A neural system is absolutely stable, when it has only asymptotically stable equilibrium points for the given weights and any input.

Tomasiello et al. (2021) have first proved the existence of fixed points of (8.33) and then that (8.33) is absolutely stable, under the assumptions that the matrix W is a diagonally dominant matrix with all entries in the range (0, 1) and the inverse of the matrix W, that is W^{-1}, is a quasi-identity matrix with unitary norm. A quasi-identity matrix with unitary norm is meant as an identity matrix with off-diagonal entries which are zero or close to zero. This assumption is reasonable, since matrix W is a block matrix, where each block is a diagonal matrix with entries in the range $(0, 1)$.

8.2.2.1 GRNN Code

The code here presented implements the GRNN with raised cosine basic functions (7.2). The function $[\texttt{rmse1}, \texttt{rmse2}] = \texttt{grnn}(\texttt{PU}, \texttt{TY}, \texttt{PUt}, \texttt{TYt}, \texttt{n}, \texttt{rp})$ returns the RMSE for training an testing (`rmse1` and `rmse2`, respectively), given input and target data matrices (`PU` and `TY`, respectively) for training, input and target data matrices (`PUt` and `TYt`, respectively) for testing, number of basic functions `n`, regularization parameter `rp`.

The main part of the code is related to the matrix of internal state, which depends on the weight matrices `W`, `WIN`, `WB`, whose entries are randomly assigned

```
W0=rand(MM,MM)*0.001;
W=eye(NN,NN)-kron(W0,eye(r,r));
Wi0=rand(P,P)*0.1;
WIN=kron(Wi0,eye(r,r));
WB=rand(NN,MM);
```

where `LL` is the number of input units, `MM` is the number of output units, `NN` is the number of inner states. Hence, the matrix of inner states `XX1` is obtained as follows, starting with a cell array of empty matrices (command `cell`):

```
XX=cell(Tmax,1);
X0=rand(NN,1);
XS=zeros(NN,1);
for i=2:Tmax
for s=1:i-1
XS=XS+W^(s-1)*(WIN*PU(i-s,:)'+WB*TY(i-s,:)');
end
XXi=(W^i*X0+XS)';
end
XX1=cell2mat(XX)
```

where `Tmax=size(PU)(1)`.

A uniform fuzzy partition is then generated to cover the inner states. The instructions are very similar to the ones for the F-transform (Chap. 7). Once the matrix A of membership grades through the fuzzy partition is obtained, the training process is completed by computing the unknown weight matrix `Wf` and then the actual output `TT`

```
K1=inv(A'*A+eye(n,n)*rp)*A';
TY1=TY(1:Tmax-1,:);
```

Table 8.1 Variables and data range for the first application example

Sensor ID	Temperature (°C)	Salinity	pH	DO (μ mol kg^{-1})
01	[11.7,16.4]	[32.1,33.7]	[7.8,8.2]	[137.4,366.4]
02	[10.8,16.1]	[33.1,33.5]	[7.7,8.1]	[101.4,270.4]

```
Wf=K1*TY1;
TT=A*Wf;
```

In the test stage, the matrix of inner states is XX1t and the matrix A1 of the membership grades are formed again but by using the test data set. The weight matrix Wf computed during the training process is now used to calculate the actual output for testing:

```
TTt=A1*Wf;
```

The complete code listing is in the Appendix.

Example 8.1 Predicting dissolved oxygen
In this example, we used the data from SeapHOx deployments from the La Jolla Kelp Forest, USA (www.bco-dmo.org/dataset-deployment/455062). The data set includes pH, dissolved oxygen (DO), temperature, and salinity. The measurements were taken by two sensors at different depth, showing a significant spatial variation of DO and pH over a 10 m increase in water depth. The measurements were registered every 15 min in a campaign over the years 2010–2013, The data range for all the variables are listed in Table 8.1.

Here, 1200 samples have been considered, of which 1000 for the training and the remainder for the test. Unlike the work by Tomasiello et al. (2021), here data are not normalized and the basic functions are not cubic B-splines, but raised cosine shaped functions. The aim of the numerical experiment is one-day-ahead prediction.

In the Scilab function grnn, it has been fixed n=60; rp=0.0001;, The results by the GRNN, as they appear in the Scilab console, are shown in Fig. 8.4. In the same figure, there is also the plot of the error (difference between target and actual value) for all the 200 test data points; such data are referred to the measurements by the first sensor.

For comparative purposes, we considered an ANN with one hidden layer with p=128 units, by using the Scilab Neural Network Module. The instructions to fix the architecture and to get the network weights matrix WA are

```
net=[m p1 o];
af = ['ann_tansig_activ','ann_purelin_activ'];
lr=0.01;
WA=ann_FFBP_lm(X',T',net,af,lr);
```

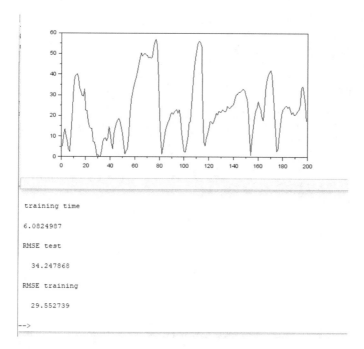

Fig. 8.4 Example 8.1: results by GRNN in the Scilab console

where X and T are the input and targets matrices for training, respectively, and m=size(X,2); o=size(T,2); are the number of input and output units, respectively.

The computed output for the testing stage is

```
Y2=ann_FFBP_run(X2',WA,af);
```

where X2 is the test input data matrix.

The results are shown in Fig. 8.5. As one can see, with regard to the training the ANN exhibits a lower RMSE ($RMSE = 16.36$), with regard to the test, the RMSE by GRNN was slightly better, i.e. $RMSE = 29.55$ against $RMSE = 33.85$ by the ANN. The training time for the GRNN is significantly lesser than the one required by the ANN, i.e. only 6 s, while for the ANN at least 149 s are needed to get the above-mentioned RMSE.

Problems

1. A fuzzy partition with small support has the additional property that there exists $r \geq 1$ such that $supp(A_i) = \{\xi \in I : A_i(\xi) > 0\} \subseteq [\xi_i, \xi_{i+r}]$. Among the possible basic functions to produce such fuzzy partitions there are B–splines. B–splines of degree p requires $m \geq p + 2$, i.e. some auxiliary points both on the left and on the right of the considered interval. An explicit form of the scaled cubic B–splines (CBS), for $j = 0, \ldots, m$, is given as follows

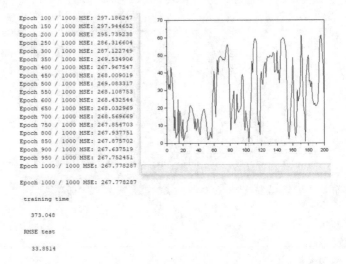

Fig. 8.5 Example 8.1: results by ANN in the Scilab console

$$
A_j(\xi) =
\begin{cases}
\dfrac{(\xi-\xi_{j-2})^3}{4h^3}, & \xi \in [\xi_{j-2}, \xi_{j-1}) \\[2mm]
\dfrac{(\xi-\xi_{j-2})^3-4(\xi-\xi_{j-1})^3}{4h^3}, & \xi \in [\xi_{j-1}, \xi_j) \\[2mm]
\dfrac{(\xi_{j+2}-\xi)^3-4(\xi_{j+1}-\xi)^3}{4h^3}, & \xi \in [\xi_j, \xi_{j+1}) \\[2mm]
\dfrac{(\xi_{j+2}-\xi)^3}{4h^3}, & \xi \in [\xi_{j+1}, \xi_{j+2}) \\[2mm]
0, & otherwise.
\end{cases}
$$

Implement a fuzzy partition using cubic B-splines.

2. Perform some numerical experiments with the GRNN using cubic B-splines and compare the results against the ones obtained by using raised cosine shaped functions.

References

Colace F, Loia V, Pedrycz W, Tomasiello S (2020) On a granular functional link network for classification. Neurocomputing 398:108–116

Zadeh LA (1997) Toward a theory of fuzzy information granulation and its centrality in human reasoning and fuzzy logic, Fuz Sets Syst, 90 (2): 111-127

Pedrycz W (2016) Granular computing - analysis and design of intelligent systems. CRC Press

Song M, Wang Y (2013) A study of granular computing in the agenda of growth of artificial neural networks. Granular Comput 1:247–257

Song M, Pedrycz W (2013) Granular neural networks: concepts and development schemes. IEEE Trans Neural Netw Learn Syst 24(4):542–553

Tomasiello S, Loia V, Khaliq A (2021) A granular recurrent neural network for multiple time series prediction. Neural Comput Appl 33:10293–10310

Appendix A
Scilab Notes and Codes

A.1 Brief Notes on Scilab and the Fuzzy Logic Toolbox

Scilab is a free open source software developed at INRIA in the 80s-90s. It covers several functional domains ranging from optimization to signal processing and systems control. The last released version is 6.1.1. The Scilab website offers many tutorials. Readers not familiar with the Scilab environment can refer to a manual for beginners available at www.scilab.org/sites/default/files/Scilab_beginners_0.pdf.

The Fuzzy Logic Toolbox (sciFLt) has been created in 2019. Details on the authors and system requirements can be found at

https://atoms.scilab.org/toolboxes/sciFLT.

To install the sciFLT, click on the **Atoms** icon in the Scilab environment and then click on **All modules** in the opened list, by finally selecting **Fuzzy Logic Toolbox**.

To start working with sciFLT, one has to load the fls Editor, by typing sciFLTEditor() in the Scilab console. Once the sciFLTEditor appears, it is possible to easily create a Mamdani or Takagi-Sugeno FIS.

To define a new FIS (Mamdami or Takagi-Sugeno), follow the path

```
File->New fls -> Takagi-Sugeno (or Mamdani)
```

For instance, for the Mamdani system (see Fig. A.1), different types of operators, implication and so on can be selected to get variants of the original FIS.

Similarly for the Takagi-Sugeno system some variants are possible. To define the input/ouput variables, one has to fix names, variables range and terms, i.e. MFs (see Figs. A.2 and A.3). In particular, trimf, trapmf, gaussmf stand for triangular, trapezoidal, gaussian MF.

For the output variable of the Takagi-Sugeno system, different linear functions can be fixed. An example is depicted in Fig. A.4. Notice that the second parameter represents the constant part. Figure A.5 shows the rule editor.

© The Editors if applicable The Author(s), under exclusive license to Springer Nature Switzerland AG 2022
S. Tomasiello et al., *Contemporary Fuzzy Logic*, Big and Integrated Artificial Intelligence 1, https://doi.org/10.1007/978-3-030-98974-3

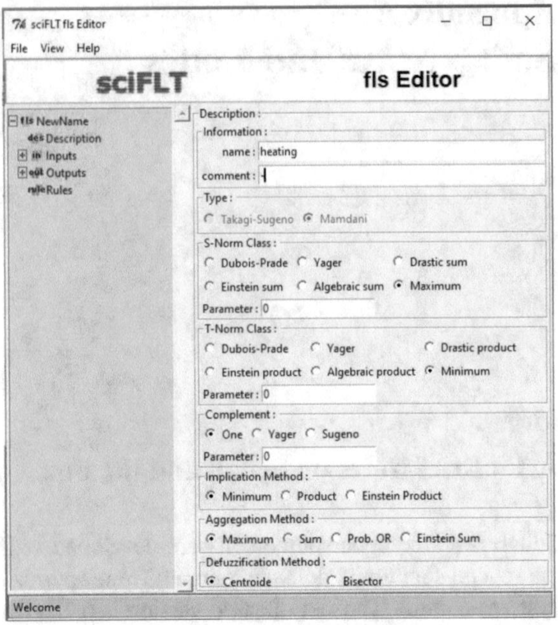

Fig. A.1 Setting general parameters in sciFLT

Fig. A.2 Setting input variables in sciFLT

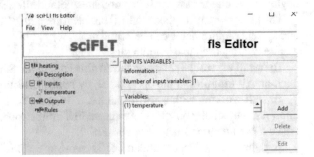

A.2   Scilab Code Listings

A.2.1   Core, Height, Support and α-Cuts

```
function [core,height,supp]=chs(fs,md)
core =fs(md==1);
height=max(md);
supp=fs(md>0);
endfunction
```

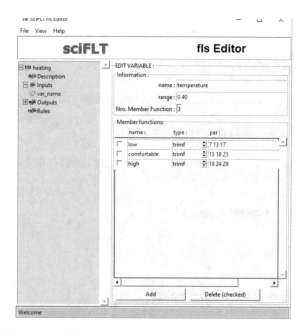

Fig. A.3 Defining MFs in sciFLT

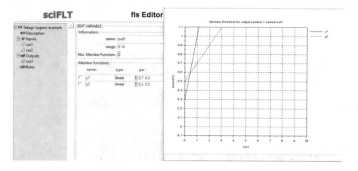

Fig. A.4 Output variable in sciFLT

Fig. A.5 Rule base editor in sciFLT

A.2.2   Intersection of Finite Fuzzy Sets

```
function ffs=fintersect(set1,md1,set2,md2)
elements = intersect(set1,set2);
if elements==[] then error('no intersection');
else
membership = zeros(1,length(elements));
for i = 1:length(elements)
membership(i) = min(md1(set1==elements(i)),md2(set2==
elements(i)));
end
ffs=[elements' membership'];
end
endfunction
```

A.2.3   Union of Finite Fuzzy Sets

```
function ffs=funion(set1,md1,set2,md2)
ss=[set1 set2];
md=[md1 md2];
fz=[ss' md'];
fz=gsort(fz,'lr','i');
inter=intersect(set1,set2);
if inter==[] then ffs=fz;
else
pp=members(fz(:,1),inter);
w=find(pp~ =0);
for i=1:2:length(w)
fz(w(i),2)=max(fz(w(i),2),fz(w(i+1),2));
end
for i=1:length(w)
fz(w(i),:)=[];
end
ffs=fz;
end
endfunction
```

A.2.4   Parametric form of Triangular Fuzzy Numbers

```
function [yl,yr]=alpha_tri(l,c,r,a)
yl=(c-l)*a+l;
yr=(c-r)*a+r;
endfunction
```

A.2.4.1   Triangular Fuzzy Number

```
function y=tri(l,c,r,x)
x1 = x(l< x & x <= c);
y(find(l< x & x <= c)) = (x1 - l)/(c-l);
x2 = x(c< x & x < r);
y(find(c< x & x < r)) = (r - x2)/(r-c);
y(find(x<=l | x>=r)) =0;
endfunction
```

A.2.5   Trapezoidal Fuzzy Number

```
function y=tra(l,b,c,r,x)
x1 = x(l< x & x <b);
y(find(l< x & x <b)) = (x1 - l)/(b-l);
x2=x(b<= x & x <=c);
y(find(b<= x & x <=c)) =1;
x3 = x(c< x & x < r);
y(find(c< x & x <r)) = (r - x3)/(r-c);
y(find(x <= l | x>=r)) =0;
endfunction
```

A.2.6   Multiplication of Triangular Fuzzy Numbers

```
function [yl,yr]=alpha_tri_m(l1,c1,r1,l2,c2,r2,a)
P=zeros(size(a,1),4);
k=1;
A=[(c1-l1)*a+l1,(c1-r1)*a+r1];
B=[(c2-l2)*a+l2,(c2-r2)*a+r2];
for i=1:2
for j=1:2
P(:,k)=A(:,j).*B(:,i);
```

```
k=k+1;
end
end
Q=gsort(P,"c","i");
yl=Q(:,1);
yr=Q(:,4);
endfunction
```

A.2.7   Division of Triangular Fuzzy Numbers

```
function [yl,yr]=alpha_tri_d(l1,c1,r1,l2,c2,r2,a)
P=zeros(size(a,1),4);
k=1;
if r2==0 |l2==0 then
a(find(a==0))=[];
else if l2<0 & r2>0 then
a(find(a==-l2/(c2-l2)))=[];
end
end
A=[(c1-l1)*a+l1,(c1-r1)*a+r1];
B=[(c2-l2)*a+l2,(c2-r2)*a+r2];
for i=1:2
for j=1:2
P(:,k)=A(:,j)./B(:,i);
k=k+1;
end
end
Q=gsort(P,"c","i");
yl=Q(:,1);
yr=Q(:,4);
plot(a,yl,a,yr);
endfunction
```

A.2.8   Max-Min Composition

```
function C = max_min(A, B)
[m,n] = size(A);
[p,q] = size(B);
if (n~= p)error('The matrices have incompatible sizes.');
end
C = zeros(m,q);
for i = 1:m
```

```
for j = 1:q
tm= min(A(i, :), B(:, j)');
C(i,j) = max(tm);
end
end
endfunction
```

A.2.9   T-norm

```
function tn=t_norm(A,B,s)
select s
case 1 then
tn=min(A,B);
case 2
tn=A.*B;
case 3 tn=max(0,(A+B-1));
else tn = zeros(A);
tn(find(B==1)) = A(find(B==1));
tn(find(A==1)) = B(find(A==1));
end
endfunction
```

A.2.10   T-conorm

```
function tn=t_conorm(A,B,s)
select s
case 1 then
tn=max(A,B);
case 2
tn=A+B-A.*B;
case 3 tn=min(1,(A+B));
else tn = ones(A);
tn(find(B==0)) = A(find(B==0));
tn(find(A==0)) = B(find(A==0));
end
endfunction
```

A.2.11   ANFIS-T

```
function [Y,rmse2]=anfist(In,Int,T,Tt,NumInTerms,C,a)
NumSamples=size(In,1);
NumSamplest=size(Int,1);
```

```
NumInVars=size(In,2);
NumRules = NumInTerms;
Out1= zeros(NumInVars,NumInTerms);
Out2= zeros(NumInVars,NumInTerms);
Out3= zeros(1,NumRules);
H0=cell(NumSamples,NumRules);
Theta = zeros((NumInVars+1)*NumRules,1);
alpha= zeros(NumInVars,NumInTerms);
sigma= zeros(NumInVars,NumInTerms);
In = In';
Int = Int';
for i=1:NumInVars
alpha(i,:) = linspace(min(In(i,:)),max(In1(i,:)),NumInTerms);
sigma(i,:) = ((alpha(i,2) - alpha(i,1))/2)*ones(1,NumInTerms);
end
In0=In;
In(NumInVars+1,:)=1;
for j=1:NumSamples
In2 = In0(:,j)*ones(1,NumInTerms);
Out1 = exp(-((In2-alpha)./sigma).^2);
Out2 = prod(Out1,1);
S_2 = sum(Out2);
if S_2~=0
Out3 = Out2 ./S_2;
else
Out3 = zeros(1,NumRules);
end
for k=1:NumRules
H0{j,k}=Out3(k)*In'(j,:);
end
end
H=cell2mat(H0);
M=H'*H;
PH=inv(eye(M)*C+M^((a+1)/2))*M^((a-1)/2)*H';
Theta=PH*T;
H0t=cell(NumSamplest,NumRules);
In0t=Int;
Int(NumInVars+1,:)=1;
for j=1:NumSamplest
In2t = In0t(:,j)*ones(1,NumInTerms);
Out1t = exp(-((In2-alpha)./sigma).^2);
Out2t = prod(Out1t,1);
S_2t = sum(Out2t);
if S_2t~=0
Out3t = Out2t./S_2t;
```

```
else
Out3t = zeros(1,NumRules);
end
for k=1:NumRules
H0t{j,k}=Out3t(k)*Int'(j,:);
end
end
H2=cell2mat(H0t);
Y=H2*Theta;
ERt=T1t-Y;
rmse2= norm(ERt)/sqrt(NumSamplest);
endfunction
```

A.2.12   F-transform

```
function [rmse,RX,F]=ft(X,n)
ss=size(X);
D=zeros(n,1);
for i=1:n
D(i)=(i - 1)/(n - 1)*(ss(1) - 1) + 1;
end
h=(D(n)-D(1))/(n-1);
B=eye(ss(2));
A=zeros(ss(1),n);
F=zeros(n,ss(2));
M1=zeros(n,ss(2));
M2=zeros(n,ss(2));
I=ones(ss(1),ss(2));
for k=1:ss(1)
if D(1)<=k & k<=D(2)
A(k,1)=0.5*(cos(%pi*(k-D(1))/h)+1);
else
if D(n-1)<=k & k<=D(n)
A(k,n)=0.5*(cos(%pi*(k-D(n))/h)+1);
end
end
for j=2:n-1
if D(j-1)<=k & k<=D(j+1)
A(k,j)=0.5*(cos(%pi*(k-D(j))/h)+1);
end
end
end
M1=A'*X*B;
```

```
M2=A'*I*B;
F=M1./M2;
RX=A*F*B';
ER=abs(X - RX);
sz=size(ER,"*");
EE=matrix(ER,sz,1)
rmse=norm(EE,2)/sqrt(sz);
end
```

A.2.13   GRNN

```
function [rmse1,rmse2]=grnn(PU,TY,PUt,TYt,n,rp)
LL=size(PU)(2);
MM=size(TY)(2);
NN=LL;
r=floor(NN/MM);
Tmax=size(PU)(1);
A=zeros(Tmax-1,n);
D=zeros(n,1);
tic();
W0=rand(MM,MM)*0.001;
W=eye(NN,NN)-kron(W0,eye(r,r));
Wi0=rand(P,P)*0.1;
WIN=kron(Wi0,eye(r,r));
WB=rand(NN,MM);
XX=cell(Tmax,1);
X0=rand(NN,1);
XS=zeros(NN,1);
for i=2:Tmax
for s=1:i-1
XS=XS+W^(s-1)*(WIN*PU(i-s,:)'+WB*TY(i-s,:)');
end
XX{i}=(W^i*X0+XS)';
end
XX1=cell2mat(XX);
for i=1:n
D(i)=(i - 1)/(n - 1)*(max(abs(XX1))-min(abs(XX1)))-min(abs(XX1));
end
h=(D(n)-D(1))/(n-1);
for k=1:Tmax-1
if D(1)<=k & k<=D(2)
A(k,1)=0.5*(cos(%pi*(k-D(1))/h)+1);
else
if D(n-1)<=k & k<=D(n)
```

```
A(k,n)=0.5*(cos(%pi*(k-D(n))/h)+1);
end
end
for j=2:n-1
if D(j-1)<=k & k<=D(j+1)
A(k,j)=0.5*(cos(%pi*(k-D(j))/h)+1);
end
end
end
K1=inv(A'*A+eye(n,n)*rp)*A';
TY1=TY(1:Tmax-1,:);
Wf=K1*TY1;
TT=A*Wf;
t=toc();
ER = abs(TT - TY1);
sz=size(ER,"*");
EE=matrix(ER,sz,1)
rmse1=norm(EE,2)/sqrt(sz);
Tmaxt=size(PUt)(1);
XXt=cell(Tmaxt,1);
A1=zeros(Tmaxt-1,n);
X0t = XX1(Tmax - 1,:)';
XSt=zeros(NN,1);
for i=2:Tmaxt
for s=1:i-1
XSt=XSt+W^(s-1)*(WIN*PUt(i-s,:)'+WB*TYt(i-s,:)');
end
XXt{i}=(W^i*X0t+XSt)';
end
XX1t=cell2mat(XXt);
for i=1:n
Dt(i)=(i - 1)/(n - 1)*(max(abs(XX1t))-min(abs(XX1t)))-min(abs(XX1t));
end
ht=(Dt(n)-Dt(1))/(n-1);
for k=1:Tmaxt-1
if Dt(1)<=k & k<=Dt(2)
A1(k,1)=0.5*(cos(%pi*(k-Dt(1))/ht)+1);
else
if Dt(n-1)<=k & k<=Dt(n)
A1(k,n)=0.5*(cos(%pi*(k-Dt(n))/ht)+1);
end
end
for j=2:n-1
if Dt(j-1)<=k & k<=Dt(j+1)
A1(k,j)=0.5*(cos(%pi*(k-Dt(j))/ht)+1);
```

```
end
end
end
TYt1=TYt(1:Tmaxt-1,:);
TTt=A1*Wf;
ERt = abs(TTt - TYt1);
szt=size(ERt,"*");
EEt=matrix(ERt,szt,1)
plot(ERt(:,1))
rmse2=norm(EEt,2)/sqrt(szt);
disp(t,'training time');
disp(rmse1,'RMSE training',rmse2, 'RMSE test');
endfunction
```

A.3   License

- It is not permitted to use any code or part of the codes herein presented in commercial and/or military applications.
- Each code is provided "as-is" without warranty of any kind, including but not limited to the implied warranties or conditions of merchantability or fitness for a particular purpose. In no events shall the authors of this book be liable for any special, incidental, indirect or consequential damages of any kind, or damages whatsoever resulting from loss of use, data, or profits, whether or not the authors have been advised of the possibility of such damages, and/or on any theory of liability arising out of or in connection with the use or performance of this software.
- A reference to this book has to be included in an article, chapter or book, where any code herein presented has been used. In particular, a reference to the paper where a code has been introduced or used for the first time should be included, i.e.

  - for the GRNN code,
    S. Tomasiello, V. Loia, A. Khaliq, A granular recurrent neural network for multiple time series prediction, Neural Comput. Applic. 33, 10293–10310 (2021)
  - for the F-transform code,
    P. Hurtik, S. Tomasiello, A review on the application of fuzzy transform in data and image compression, Soft Comput. 23(23), 12641–12653 (2019)
  - for the ANFIS-T code,
    Tomasiello, S, Pedrycz, W, Loia, V, (2022) On Fractional Tikhonov Regularization: Application to the Adaptive Network-Based Fuzzy Inference System for Regression Problems. IEEE Transactions on Fuzzy Systems, DOI: 10.1109/TFUZZ.2022.3157947

Printed in the United States
by Baker & Taylor Publisher Services